Innovative Approaches in Plant Protection Under Changing Climatic Scenarios

About the Editors

Devesh Pathak, is currently worked as Assistant Professor (Plant Pathology) in RSM College, Dhampur Bijnor. He completed his graduation from RBS College, Bichpuri, Agra and post-graduation and Ph.D (Pursing) from Aligarh Muslim University, Aligarh. He also qualified ASRB-NET and his research focuses on management of various fungal diseases of pulses and oilseed crops. He is a lifetime member of various reputed societies/organizations. He has also published 11 research papers, 3 book chapters with ISBN, and 5 popular articles in national and international peer-reviewed journals and magazines. He is a recipient of 4 awards from various national and international conferences/ seminars. He also attended more than 10 national and international conferences/ seminars and training workshops.

Ravi Dhanker, is currently worked as Assistant Professor (Soil science and Ag. Chemistry) in RSM College, Dhampur, Bijnor. He completed his graduation from Amar Singh PG college, Lakhaoti, Bulandshahr and post-graduation and Ph.D. from CSAUAS&T, Kanpur. His research focuses on remote sensing and GIS technology. He has also published 15 research papers, 5 book chapters with ISBN, and 3 popular articles in national and international peer-reviewed journals and magazines. He is a recipient of 2 awards from various national and international conferences/seminars. He also attended more than 5 national and international conferences/seminars and training workshops.

Himanshu Tripathi, is currently worked as Assistant Professor (Ag. Engineering) in RSM College, Dhampur, Bijnor. He completed his graduation from AAIDU, Allahabad and M. Tech and Ph.D. SHUATS, Allahabad from the same university. He also qualified ASRB-NET and his research focuses on precision farm technology, farm mechanization, and conservation agriculture. He is a lifetime member of various reputed societies/organizations. He has also published 15 research papers, 5 book chapters with ISBN, and 45 popular articles in national and international peer-reviewed journals and magazines. He is a recipient of 5 awards from different national and international conferences/ seminars. He also attended more than 5 national and international conferences/ seminars and training workshops.

Vaibhav Pratap Singh, is currently pursuing Ph.D. in Plant Pathology from Aligarh Muslim University, Aligarh. He completed his graduation and post-graduation from the same university. He also qualified ICAR-NET and his research focuses on management of various fungal diseases of oilseed crops and pulses. He is a lifetime member of various reputed societies/organizations. He has also authored 1 book and published 12 research papers, 7 book chapters with ISBN, and 7 popular articles in national and international peer-reviewed journals and magazines. He is a recipient of 4 awards from different national and international conferences/seminars. He also attended more than 14 national and international conferences/seminars and training workshops.

1168, Sector 13, Urban Estate
Karnal-132 001, Haryana
Tel: 91-84470 75807, 18440 41168
Email: contentvibesppa@gmail.com
www.contentvibes.in

Print ISBN: 978-93-67554-39-5

ebook ISBN: 978-93-67551-07-3

Innovative Approaches in Plant Protection Under Changing Climatic Scenarios

Devesh Pathak
Assistant Professor
Department of Plant Pathology
RSM College, Dhampur
Bijnor, Uttar Pradesh

Ravi Dhanker
Assistant Professor
Department of Soil Science and Agricultural Chemistry
RSM College, Dhampur
Bijnor, Uttar Pradesh

Himanshu Tripathi
Assistant Professor
Department of Agricultural Engineering
RSM College, Dhampur
Bijnor, Uttar Pradesh

Vaibhav Pratap Singh
Research Scholar
Department of Plant Protection
Aligarh Muslim University
Aligarh, Uttar Pradesh

Preface

The rapid pace of climate change has introduced unprecedented challenges to agricultural productivity, especially in the realm of plant protection. Shifting weather patterns, rising temperatures, altered rainfall regimes, and increased frequency of extreme weather events are reshaping the dynamics of plant pests, diseases, and the effectiveness of existing pest management strategies. In this context, innovative approaches to plant protection have become essential to safeguard global food security and promote sustainable agricultural practices.

This book, ***"Innovative Approaches in Plant Protection under Changing Climatic Scenarios"***, delves into the emerging trends and advanced technologies that are transforming the field of plant protection. From genetic resistance and integrated pest management (IPM) strategies to the use of bio pesticides, smart sensors, and climate models, the chapters presented here explore how scientific breakthroughs are enabling the agricultural sector to adapt to the challenges posed by climate change.

Experts from diverse fields including plant pathology, entomology, and environmental sciences etc. offer a comprehensive view of both the threats and solutions that shape the future of plant protection. Additionally, the book discusses the potential of climate-resilient crops, the role of artificial intelligence and machine learning in pest prediction and management, and the integration of indigenous knowledge with modern technology.

As we face the reality of a rapidly changing climate, the need for adaptive, sustainable, and forward-thinking strategies in plant protection is more critical than ever. The following chapters aim to inspire researchers, policymakers, and practitioners to embrace these innovations, ensuring the protection of crops and the resilience of agricultural systems in the years to come.

Editors

Contents

Preface .... *vii*

1. **Green Pesticides: A Safe and Eco-friendly Method to Mitigate Pest and Diseases..1**
   Vaibhav Pratap Singh, Sibte Sayyeda, Devesh Pathak and Deepak Mourya
2. **Innovative Plant Protection Strategies in Horticultural Crops.....13**
   Amit Kumar, Harish Kumar, Nikhil, Saklain Mulani and Devesh Attri
3. **Geospatial Vigilance: A Crop Health Assessment Tool to Forecast Disease Outbreaks in Sustainable Agriculture 41**
   Sibte Sayyeda, Vaibhav Pratap Singh, Devesh Pathak and Deepak Mourya

4 **Improvement of Major Cereal Crops for Changing Disease and Insect Pest Scenario in India 57**
   Satyendra, Riya Raj, Sreejaya Ghosh and Prerna Suman

5. **Use of GIS Based Technology for Plant Disease and Pest Management..91**
   Satish Kumar Singh, Anurag Patel and Himanshu Tripathi
6. **Nano-Tech Approaches in Plant Disease Management 101**
   Rajesh Kumari, Devesh Pathak and Ravi Dhanker
7. **Geo-spatial Technology for Pest and Diseases Management 113**
   Mahesh Tripathi and Himanshu Tripathi
8. **Effect of Climate Change on Insect Pests and Food Security 121**
   Surendra Singh Shekhawat and Meenakshi Devi
9. **Soil Health Management: Innovative Tools for Plant Protection...137**
   Ravi Dhanker and Mukesh Kumar

10. Sustainable Management of Plant Diseases 145

Sachin Kumar Jain, Ajay Kumar, Abhishek Singh and Dushyant Kumar

11. Role of Sustainable Approaches in Management of Chickpea Pod Borer (*Helicoverpa armigera* Hubner) Population in Chickpea Crop 169

K.K. Maurya

Index...189

1

Green Pesticides A Safe and Eco-friendly Method to Mitigate Pest and Diseases

Vaibhav Pratap Singh[1], Sibte Sayyeda[1], Devesh Pathak[2] and Deepak Mourya[3]

[1]Department of Plant Protection, Faculty of Agricultural Sciences Aligarh Muslim University, Aligarh, Uttar Pradesh-202002
[2]Department of Plant Pathology, RSM College, Dhampur, Bijnor Uttar Pradesh-246761
[3]Department of Plant Pathology, Shri Khushal Das University Hanumangarh, Rajasthan-335801

Abstract

The use of conventional pesticides has raised significant concerns due to their adverse effects on human health and the environment. In response, there has been a growing interest in developing safer and more eco-friendly alternatives. This chapter explores the concept of green pesticides as a sustainable solution to pest and disease management. Green pesticides, derived from natural sources or synthesized using eco-friendly methods, offer effective control of pests while minimizing harm to non-target organisms and ecosystems. Green pesticides encompass a diverse range of products and approaches, including botanical extracts, microbial agents, natural predators, and pheromones. Moreover, there is a need to incorporate different cultural practices along with green pesticides to enhance pest control while reducing reliance on chemical inputs. However, the efficacy of green pesticides depend on several factors, including pest species, crop type, environmental conditions, and regulatory considerations. Therefore, ongoing research and innovation are essential to optimize the performance and scalability of these eco-friendly alternatives. Overall, this chapter underscores the potential of green pesticides as a safe and sustainable approach to pest and disease management in sustainable agriculture.

Keywords: *Green pesticides, Sustainable agriculture, Eco-friendly method, Pest and disease management*

Introduction

Agriculture plays a vital role in the Indian economy and accounting for about 20.2% of the country's gross domestic product (GDP). Around 70% of Indian people depend on agriculture for its livelihood (Packiam, 2018). After the green evolution, there has been a notable increment in crop produce due to application of nitrogenous fertilizers and chemical pesticides which leads to the emergence of resistance in pests and pathogens. As a result, insect pests cause significant damage to a vast amount of farm food products. To combat these destructive pests and diseases, farmers resort to using large quantities of chemical pesticides. Indiscriminate use of chemical pesticides has adverse effects on both beneficial insects and human health. To mitigate these harmful impacts, green insecticides, botanical insecticides, or plant-based alternatives are suggested as safer options compared to modern synthetic insecticides.

Impact of Chemical Pesticides on Agriculture and Human Health

The application of chemical pesticides has indeed reduced the risk of pest infestation due to their rapid knock-down effect. However, a concerning aspect is that only a small amount, approximately 0.1%, of the agrochemicals applied for crop protection actually reaches the intended target pests. The remaining 99.9% disperses into the environment, posing hazards to non-target organisms (Packiam, 2018). The widespread and indiscriminate use of pesticides over time has been shown to have detrimental effects on various aspects of the environment and human health. These effects include harm to soil micro flora, wildlife, and human life. Additionally, pesticide misuse has led to several adverse outcomes, such as the development of insect resistance, toxicity to non-target organisms, the emergence of new pests, excessive residues on seeds, vegetables, and fruits, and alterations in pest population dynamics (Saahil, 2023). These cumulative impacts contribute to soil depletion and pose significant challenges to sustainable agricultural practices. Every year, approximately one million people suffer from pesticide poisoning, often due to increased dosages and frequent applications. These alarming statistics underscore the urgent need for more environmentally friendly and safer alternatives for crop protection.

Green pesticides: A Safer and Effective Method

The term "Green pesticides" comprises a variety of eco-friendly pest control methods aimed at reducing pest populations and enhancing food production. In concept of green pesticides, several efforts have been made to incorporate various substances such as plant extracts, hormones, pheromones, and organic toxins. This approach covers various pest control methods like microbial agents, pesticides derived from plants, secondary metabolites from microorganisms, entomophagous nematodes, pheromones, and genetic modifications in crops to confer resistance against pests. Plant-based materials offer a promising alternative for crop protection as they are generally safer and effective against various pests, diseases, nematodes, and other organisms. Plants and insects have evolved over millions of years, developing specific secondary metabolites to counteract insect damage. These bioactive compounds serve various functions such as acting as insecticides, anti-feedants, insect growth regulators, juvenile hormones, ecdysones, repellents, attractants, and arrestants (Figure 1).

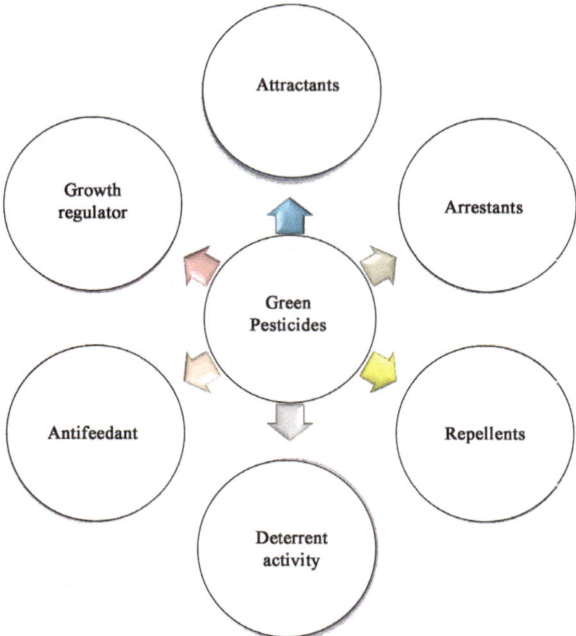

Fig. 1: Characteristics of green pesticides

Due to the diverse activities of these plant molecules, green pesticides derived from them hold significant potential as alternatives to chemical pesticides. Their multifaceted nature allows them to target pests while minimizing harm to non-target organisms and the environment. Additionally, utilizing plant-derived materials for pest control aligns with sustainable agriculture practices and reduces reliance on synthetic chemicals, thus contributing to safer and more eco-friendly crop protection methods (Koul *et al.*, 2008).

Type of Green Pesticides

1. Natural Pesticides

Living organisms utilized to manage pests are termed as biological controls or biological agents. When a microorganism undergoes packaging and marketing for pest control purposes, it is legally classified as a biopesticide and is subject to regulation accordingly (EPA, 2013). To ensure the effectiveness of biological insecticides, it's crucial to adhere closely to the instructions provided on the label. This helps prevent unintentional harm to the organisms within the product, thus maintaining its efficacy. However, it's important to note that even though they're natural, they should still be used carefully and according to label instructions to minimize any potential risks to humans, animals, or the environment.

(a) ***Bacillus thuringiensis* (Bt):** It is a naturally occurring bacterium that targets the larval stages of insect pests like mosquitoes, Colorado potato beetles, and cabbage loopers. *Bacillus thuringiensis* var. *kurstaki* (Bt), targets Lepidopteran larvae like caterpillars and are commonly found on vegetables and fruits. In environment, when a caterpillar ingested Bt, the bacterium releases a toxin inside the insect's gut. This toxin then degrades the stomach lining of the caterpillar, ultimately leading to the insect's death. Commercial Bt products consist of the bacterial toxin and are nonliving. Bt's effectiveness can be affected by ultraviolet light (sunlight), making it most efficient when applied during overcast conditions or late in the day. Typically, Bt products degrade within 24 hours, irrespective of sunlight exposure or temperature variations, resulting in a brief period of effectiveness following application (Nnamonu and Onekutu, 2015).

(b) ***Beauveria bassiana:*** This soil-borne fungus is effective for controlling various pests such as thrips, aphids, whitefly, caterpillars, beetles, and subterranean insects like ants and termites. It is applied to target as spores (reproductive structure of fungus). The fungal hyphae penetrate the insect, consuming its body tissue, generating toxins, and reproducing within it (Nnamonu and Onekutu, 2015). Typically, it takes up to seven days for the insect death. *B. bassiana* can be sensitive to sunlight, hence it's advisable to apply it towards the end of the day. However, one major drawback of this insecticide is that its spores can infect numerous non-target beneficial insects as well.

(c) **Nematodes:** Nematodes are multicellular organisms, commonly known as round worms. Cruising nematodes actively seek out their insect prey, whereas ambushing nematodes wait for insects to pass by. *Steinernema carpocapsae* and *S. scapterisici* (ambushing category) are effective for controlling mobile insects such as crane fly larvae, fleas, and cutworms. On the other hand, *Heterorhabditis bacteriophora* and *S. glaseri*, categorized as cruisers, are used to manage slower-moving insects like root weevil and scarab beetle larvae. Some nematode species, such as *S. riobrave* and *S. feltiae*, exhibit both ambushing and cruising behaviors, making them effective at managing pests like codling moth and fruit tree borers. The nematodes release bacteria that liquefy the innards of the insect. This process turns the insect into food for the nematodes, allowing them to multiply.

(d) ***Nosema*:** Nosema, a protozoan, is found to be effective in controlling some insect pests. *Nosema locustae* is employed to control grasshopper populations. Its spores are incorporated into bait, also known as an attractant, which is consumed by the grasshoppers. Once ingested, the spores germinate, and the protozoans thrive within the insect's body cavity.

2. Botanical Pesticides

The historical use of plant-derived compounds as botanical insecticides dates back thousands of years, with evidence dating to around 400 B.C. Extracts from chrysanthemum species were utilized to manage insect pests (Silva-Aguayo, 2009). Botanical insecticides are typically obtained through a process involving the maceration (soaking and separating) of plant tissues containing high concentrations of the active ingredient, followed by distillation (evaporation and condensation) to isolate the specific active compounds. The benefit of using botanical insecticides lies in their short persistence and minimal bioaccumulation in the environment, attributed to rapid degradation. Studies conducted on various experimental botanical pesticides, such as piperamides and alpha-terthienyl, indicate that they degrade within hours or days after application (Nnamonu and Onekutu, 2015). Nevertheless, the short persistence of botanical insecticides can also be considered a disadvantage, as it may necessitate multiple applications to achieve sufficient pest suppression.

While botanical pesticides provide several advantages, still the market faces several significant challenges. Despite experiencing some growth, it has not expanded at a rate comparable to the botanical medicine market in recent years. The primary obstacle lies in the expensive toxicology testing required for new products, especially when they offer limited Intellectual Property (IP) protection and target a relatively small market. Additional challenges include ensuring an economical supply of plant products, maintaining quality control, and addressing issues related to stability. Moreover, there is competition from other biopesticides and bio-control agents in the market.

(a). Neem: The neem tree is native to India and serves as the source of numerous products, including insecticides derived from extracts of the seeds and bark. The active ingredient is azadirachtin. When azadirachtin is employed for pest management, it exhibits multiple functions such as acting as an insect repellent, an anti-feedant (disrupting feeding), and a growth regulator (impairing molting and growth). Azadirachtin has been associated with adverse effects on the gastrointestinal tract of insects, affecting processes such as enzyme production, cell proliferation, and motility (Timmins and Reynolds, 1992; Nasiruddin and Mordue (Luntz), 1993a). In the case of neem oil or neem soap application, it functions similarly to other soaps and oils by poisoning upon contact. Neem insecticides prove effectiveness against a variety of pests, including caterpillars, flies, whitefly, and scales. They also exhibit some efficacy against aphids, although to a lesser extent. Achieving effective management of targeted pests often requires multiple applications of neem insecticides. Neem is considered non-toxic to vertebrate animals and has been observed to have minimal impact on many beneficial insects, including bees, spiders, and ladybugs.

Table 1: List of some commercial biopesticides and their formulations

Commercial name	**Active Ingredients and Formulation**	**Manufacturer**
Neem Aura	Neem leaf extract, Aloe vera, extract of barberry, camomile, goldenseal, myrrh, neem, and thyme; oil of anise, cedarwood, citronella, coconut, lavender, lemongrass, neem, orange	Neem Aura Naturals, Alachua, FL
GonE!	Aloe vera, camphor, menthol, oils of eucalyptus, lavender, rosemary, sage, and soybean	Aubrey Organics, Tampa, FL
MosquitoSafe	Geraniol 25%, mineral oil 74%, Aloe vera 1%	Naturale, Ltd., Great Neck, NY
Repel	Lemon eucalyptus insect repellent lotion. Oil of lemon eucalyptus (65% p-menthane-3,8-diol [PMD]) (26%)	Wisconsin Pharmacol Comp., Inc., Jackson, WI
Natrapel	Citronella (10%)	Tender Corp., Littleton, NH
SunSwat	Oils of bay, cedarwood, citronella, goldenseal, juniper, lavender, lemon peel, patchouli, pennyroyal, tansy,tea tree, and vetivert	Kiss My Face Corp., Gardiner, NY

Commercial name	Active Ingredients and Formulation	Manufacturer
Bygone	Oils of canola, eucalyptus, peppermint, rosemary, and sweet birch,	Lakon Herbals, Inc., Montpelier, VT
Skinsations	Deet (N,N-diethyl-3-methylbenzamide, 7%),	Spectrum Corp., St. Louis, MO Spray
Nimbecidine	Neem seed kernel extracts	Agriplex Pvt. Ltd. Bengaluru, India

Source: (Kaur and Kaur, 2017)

(b) Pyrethrum: It is derived from the seeds of *Chrysanthemum cinerariaefolium* and has served as an insecticide for over a century. Pyrethrum demonstrates effectiveness against various soft-bodied garden pests like scales, whitefly, mealybugs, and thrips, yet it does not effectively control mites. Pyrethrins, found in pyrethrum, act as neurotoxins by targeting the nervous system of insects and leading to repeated and prolonged nerve firings. Additionally, they may exhibit a repellent effect. Pyrethrum serves as a broad-spectrum insecticide with toxicity extending to beneficial insects. Upon exposure, it can induce paralysis in susceptible insects; however, it degrades rapidly in sunlight, typically within hours. Consequently, repeated applications are often necessary to achieve adequate pest management. Pyrethrum formulations commonly incorporate low-hazard activators or synergists, like piperonyl butoxide or piperonyl cyclonene, to enhance effectiveness significantly and lower production costs (Pedigo and Rice, 2008).

(c). Horticultural oil: Horticultural oils were used for insect control as early as 1763, and they remain a popular choice for today. While many of these control agents are petroleum- based, there are also plant-based oils available that are considered acceptable in organic farming practices. Horticultural oils function by disrupting insect feeding and egg-laying processes when the pest is completely coated. Eggs covered with oils are unable to exchange gases, leading to the suffocation of the developing pest. When used properly, horticultural oils have minimal phytotoxic effects on plants. However, the timing of application, plant species, temperature, and type of oil all play roles in determining effectiveness and the risk of phytotoxicity. Phytotoxic effects are easily observed and characterized by browning or "burning" of leaves or new growth on stems (Nnamonu and Onekutu, 2015).

Other traditionally used botanical insecticide products include nicotine, rotenone, ryania, sabadilla, and pyrethrum. Rotenone, derived from extracts of tropical legumes *Derris* and *Lonchocarpus*, serves as a trade name for the insecticide. Its main active principle, the isoflavonoid rotenone, possesses moderate toxicity to mammals due to poor absorption and rapid metabolism. However, it proves highly toxic to insects and fish, owing to its rapid uptake and inhibition of respiratory electron transport. Rotenone has been widely used as a plant extract at reasonable costs and utilized as a dust on horticultural and ornamental crops. Sabadilla is derived from the seed extract of the neo-tropical lily *Schoenocaulon officinale*, containing veratridine alkaloids that operate through a neurotoxic mode of action. This extract exhibits low mammalian toxicity and serves as a valuable contact insecticide against various agricultural insects, including lepidoptera, leafhoppers, and thrips. Ryania, on the other hand, originates from the South American shrub *Ryania* sp. It contains the diterpene alkaloid ryanodine and functions as a contact and ingested insecticide targeting horticultural and ornamental crop pests.

3. Essential Oil as Green Pesticides

Essential oils are characterized as volatile oils with potent aromatic components, imparting distinct odors, flavors, or scents to plants. They arise as by-products of plant metabolism and are often referred to as volatile plant secondary metabolites. These oils are typically located in glandular hairs or secretory cavities within plant cell walls, manifesting as fluid droplets present in various plant parts such as leaves, stems, bark, flowers, roots, and fruits (Lawrence and Reynolds, 2001). The aromatic properties of essential oils serve several functions for plants, including (i) attracting or repelling insects, (ii) shielding themselves from heat or cold, and (iii) employing chemical constituents within the oil as defense mechanisms. The majority of essential oils consist of monoterpenes, which are compounds containing 10 carbon atoms often arranged in either a ring or acyclic form. Additionally, sesquiterpenes, hydrocarbons comprising 15 carbon atoms, are commonly found in essential oils. Higher terpenes may also be present as minor constituents. The most prevalent groups within essential oils are cyclic compounds featuring saturated or unsaturated hexacyclic or aromatic systems. Examples of bicyclic compounds include 1,8-cineole, while acyclic examples encompass linalool and citronellal.

Numerous essential oils like rose, patchouli, sandalwood, lavender, and geranium have established reputations in the perfumery and fragrance industry. Conversely, essential oils like lemongrass, eucalyptus, rosemary, vetiver, clove, and thyme are renowned for their pest control properties. Peppermint acts as a repellent against ants, flies, lice, and moths, while pennyroyal deters fleas, ants, lice, mosquitoes, ticks, and moths. Additionally, spearmint (*Mentha spicata*) and basil (*Ocimum basilicum*) have proven effective in repelling flies. Essential oils extracted from eucalyptus and lemongrass have demonstrated efficacy as animal repellents, antifeedants, miticides, insecticides and antimicrobial agents. Consequently, they have been utilized as disinfectants, sanitizers, bacteriostats, microbiocides, and fungicides, making significant contributions to safeguarding household possessions. Citronella essential oil, derived from *Cymbopogon nardus*, has been utilized for more than fifty years as both an insect repellent and an animal repellent. A combination of a few drops each of citronella, lemon (*Citrus limon*), rose (*Rosa damascena*), lavender, and basil essential oils mixed with one liter of distilled water has proven effective in repelling indoor insect pests (Zaridah *et al.*, 2003).

Merits and Demerits of Green pesticides

Green pesticides are safe, cost-effective and eco-friendly in nature. However, like many benefits, they also have some drawbacks.

Merits

1. **Cost-effectiveness:** Green pesticides are often cheaper compared to synthetic alternatives, making them accessible to farmers with limited budgets.
2. **Environmental friendly:** These pesticides are economically viable and easily biodegradable in nature because these pesticides are derived from natural sources, reducing the overall ecological impact compared to chemical pesticides.
3. **Effectiveness:** Despite being natural, many green pesticides are still effective in controlling pests and diseases, providing reliable pest management solutions.

Demerits

1. **Limited effectiveness:** Some green pesticides may not be as potent as synthetic chemicals, leading to less effective pest control in certain situations.
2. **Shorter shelf life:** Green pesticides, being derived from natural sources, may have a shorter shelf life compared to synthetic alternatives, requiring more frequent applications.
3. **Residue and persistence:** While generally safer for the environment, some green pesticides may still leave residues on crops or persist in the environment, posing potential risks to non-target organisms.
4. **Variable efficacy:** The effectiveness of green pesticides can vary depending on factors such as application method, pest type, and environmental conditions, leading to inconsistent results.

Conclusion

The use of green pesticides as a safe and eco-friendly method to mitigate pests and diseases reveals a promising pathway towards sustainable agriculture. The conventional reliance on synthetic pesticides has led to significant environmental and health concerns, highlighting the urgent need for alternative approaches. Green pesticides include botanical extracts, microbial agents, natural predators, and pheromones which offer viable alternatives that prioritize environmental stewardship without compromising efficacy. However, the adoption of green pesticides is not without challenges. Various factors such as pest species, crop type, environmental conditions, and regulatory frameworks can influence their efficacy and practicality. Ongoing research and innovation are therefore crucial to optimize the performance and scalability of these eco-friendly alternatives. By promoting biodiversity, minimizing chemical inputs, and fostering resilience in agricultural systems, they contribute to build a more sustainable and resilient food production system. Moreover, their adoption aligns with consumer preferences for safer and more environmentally friendly products. Green pesticides represent a paradigm shift in pest management and offering a pathway towards a more sustainable agricultural future. By integrating these eco-friendly alternatives into holistic pest management strategies, we can protect crops, safeguard ecosystems, and ensure food security for generations to come.

References

EPA (2013). US Environment Protection Agency, USA (https://www.epa.gov).

Kaur, A. and Kaur, P. (2017). Green pesticides for clean environment. International Journal of Engineering Development and Research. Volume 5, Issue 4, 1347-1349.

Koul, O., Walia, S. and Dhaliwal, G. S. (2008), Essential Oils as Green Pesticides: Potential and Constraints, Biopestic. Int., 4(1): 63-84.

Lawrence, B. M. & Reynolds, R. J. (2001), Progress in essential oils. Perf. Flavour., 26, 44–52.

Nasiruddin M., Mordue Luntz A. J. (1993). The effect of azadirachtin on the midgut histology of the locusts, Schistocerca gregaria and Locusta migratoria. Tissue Cell, 25, 875-884.

Nnamonu, L. A. and Onekutu, A. (2015). Green Pesticides in Nigeria: An Overview, Journal of Biology, Agriculture and Healthcare, Vol.5, No.9, 48-62.

Packiam, M. S. (2018). Green Pesticides: Eco-friendly Technology for Integrated Pest Management. Acta Scientific Agriculture, 2, 11: 01.

Parkash, A., Rao, J. and Nandagopal, V. (2008). Future of botanical pesticides in rice, wheat, pulses and vegetable pest management. Journal of Biopesticides. 1: 154-169.

Pedigo, L. P. and M. E. Rice. (2009). Entomology and pest management. New Jersey: Prentice Hall.

Saahil (2023). Green Pesticide- An Eco-Friendly Approach for Pest Management. Just Agriculture multidisciplinary e-newsletter. Vol. 3, Issue-10, 226-230.

Silva-Aguayo, G. Botanical insecticides. Chapter in Radcliffe's 1PM World Textbook. Simpson, B. B., and Ogorzaly, M. C. (1995). Economic botany: Plants in our world. (2nd ed.). New York: McGraw—Hill. Stewart, A.

Timmins W., Reynolds S. (1992). Azadirachtin inhibits secretion of trypsin in midgut of Manduca sexta caterpillars: reduced growth due to impaired protein digestion. Entomol. Exp. Appl., 63, 47-54.

Zaridah, M.Z., Nor Azah, M.A., Abu Said, A. & Mohd. Faridz, Z.P. (2003) Larvicidal properties of citronellal and Cymbopogon nardus essential oils from two different localities. Trop. Biomed., 20, 169-174.

2

Innovative Plant Protection Strategies in Horticultural Crops

Amit Kumar[1*], Harish Kumar[2], Nikhil[3], Saklain Mulani[1] and Devesh Attri[1]

[1]Division of Fruit Science, SKUAST-Kashmir, Shalimar Campus Srinagar, Jammu and Kashmir, India
[2]Division of Vegetable Science, SKUAST-Kashmir, Shalimar Campus Srinagar, Jammu and Kashmir, India
[3]School of Agricultural, SKUAST-Kashmir, Shalimar Campus Srinagar, Jammu and Kashmir, India

Plant disease causes losses estimated for 20-40 per cent of global crop. Climate change has a very complex effect on plant-pathogen interactions, since environmental conditions affect the whole disease triangle: they modify plant susceptibility, the biological cycles of parasites and pathogens, and host-pathogen physiology and interactions. Innovative plant protection strategies in horticultural crops encompass a wide range of techniques and approaches aimed at safeguarding plants from pests, diseases, and environmental stresses in a sustainable and effective manner. These strategies are essential for ensuring the productivity, quality, and resilience of horticultural crops while minimizing negative impacts on the environment and human health.

Climate change, including increased temperature, CO_2 levels, acid rain, and tropospheric O_3, stresses plants and lowers their pathogen response. While higher CO_2 reduces stomata density, acid rain and O_3 weaken cuticle protection, aiding pathogen entry. Climate change also increases pesticide use due to longer growing seasons and altered pesticide dynamics. Intense rainfall can wash off fungicides, reducing control. Elevated CO_2 affects pesticide uptake and canopy density, hindering spray coverage and increasing pathogen resistance. Microbial biological control agents are crucial to mitigate the environmental, social, and economic impacts of increased pesticide use. These agents enhance plant disease resistance through systemic acquired resistance (SAR) and induced systemic resistance (ISR). SAR involves salicylic acid accumulation and immune response activation, while ISR, induced by beneficial microorganisms, primes the plant immune system for a rapid response without affecting growth and yield.

Role of Microbial Bio-Stimulants in Response to Biotic-Stresses

The protection against biotic stresses falls outside the generally accepted definition of bio-stimulation, disease resistance induction is sometimes elicited by stricto sensu bio- stimulants and will be discussed here as a desirable additional trait for future products to be developed. The microorganisms directly act on pests and pathogens (such as entomopathogenic and antibiotic-producing microbes), which are categorized as bio-pesticides rather than bio-stimulants. For instance, the downy mildew of Grapevine, graymold of potato and bacterial canker of kiwifruits represent crop diseases whose incidence has been increased due to climate change. The presence of even more frequent frost events during the vegetative season led to an increase in disease incidence of the bacterial canker of kiwifruit (*Pseudomonas syringae* pv. *actinidiae*), as plant tissues were more subjected to frost damages, which can be exploited by the pathogen as entry points.

Bacterial species of the genera *Pseudomonas*, *Serratia* and *Bacillus*, and fungi such as *Trichoderma* spp. and *Piriformospora indica*, are among the most studied organisms for the induction of resistance. The commercial formulation Trianum-P® (Koppert, Srl, Verona, Italy), for instance, employs a *Trichoderma harzianum* isolate to induce ISR and protection against soil pathogens. In *Pseudomonas* and *Bacillus* species. Based products, resistance induction effect may as well exist, although not documented. Several microbial molecules can activate ISR, such as flagellar proteins, Gram-negative bacteria lipopolysaccharides, siderophores, some antibiotics, N-alkylated benzylamines, VOCs and N-acyl homoserine lactones, a class of signal molecules involved in bacterial quorum sensing. The antifungal compound 2,4-diacetylfloroglucinol and cyclic lipopeptides are also recognized as microbial elicitors. These elicitors stimulate plant immune response via the activation of the regulatory genes involved in ethylene and jasmonic acid biosynthesis *Enterobacter asburiae* R57 strain, recently isolated from raspberry, showed the ability to control *Botritys cinerea in vitro* via the production of siderophores and acetoin, a volatile precursor of 2,3-butanediol.

Other microorganisms can be beneficial for plant resistance, increasing plant constitutive barriers, for example by promoting callose deposition in cell wall following ABA stimulation Even preventive treatments with ACC deaminase-producing bacteria could help in protecting plants from bacteria, fungi and nematodes, impeding the development of symptoms and decreasing disease severity. This kind of response was detected following the application of the ACC deaminase producing *Pseudomonas putida* UW4, which limited damages caused by *Pythium ultimum* in cucumber. Several microbes are registered as active principles of pesticides, acting against plant pathogens through direct antagonism mechanisms. In contrast, broad sense bio-stimulation (*i.e.,* resistance induction) properties against biotic stresses are less considered for commercial formulates. *Bacillus amyloliquefaciens* (formerly subtilis) QST 713 is an EU registered pesticide active substance that, besides directly competing for nutrients on leaf surfaces with fungal pathogens, induces systemic resistance responses in plants, as indicated by peroxidase production. Among complex commercial products, Ryze® (L. Gobbi, Srl, Genova, Italy) and Nutribac® (Chemia, Spa, Ferrara, Italy), containing both mycorrhizae and PGPR, exhibit not only general beneficial properties for the soil, but also enhance plant resistance to biotic stresses. Direct antagonism seems to have been a more appealing strategy for microbe-based products commercialization. Alternatively, resistance-inducing products (such as chitosan, exopolysaccharides and lipopolysaccharides) have been isolated and extracted from their originating organism. While the use of live microbes as resistance inducers may be a less straightforward, thus, possibly less reliable protection strategy, it may as well combine the advantages of a long-lasting and stable plant-microbe interaction and of wide-range protection, possibly improving the baseline health status of crops and reducing the synergism within pathogen consortia

Limitations and Future Perspectives in the Applications of Microbial Bio-Stimulants

The development of a new microbial biostimulant displays some specific difficulties. Firstly, commercial registration process is usually complex, and a harmonized international legislation is still lacking. Secondly, product efficacy is strictly dependent on the horticultural crop on which it is applied and on its phenological state. The development of phyto-stimulant products needs to evaluate the relationship microorganisms establish with the host plant. A positive and long-lasting plant colonization is an essential prerequisite for biostimulant effectiveness. Finally, the best formulation to guarantee products efficacy and conservation has to be defined, to minimize the influences from environmental and cultural conditions. Biostimulants with Gram-positive bacteria allow powder formulations with durable stability and drying tolerance due to spore production by bacteria. Even if there are many examples of the efficacy of microorganism application in promoting plant growth under unfavourable conditions, very few biostimulant products specifically addressing stresses emphasized by climate change are available. In general, the high costs related to the production of the commercial biostimulant and the variability in the efficacy observed in field conditions are major hindrances to the development of biostimulant products, resulting in a relatively low number of commercialized products and a limited diffusion in horticultural practice. Several parameters in fact have to be considered before the application of a microbial biostimulant:

1. Soil and crop characteristics: no microorganisms can be universally applied in any ecosystem or on any vegetable host, thus, choosing a particular strain for a biostimulant product needs consideration of soil properties and specific crop requirements, in order to select microbial strains with the best adaptation to each particular condition
2. Competition for nutrients and ecological niche occupation between selected microbial strains and indigenous microflora, which can reduce biostimulant efficacy.
3. Mode of application of the microbial biostimulant, that should reduce micro-organisms dispersion or death due to abiotic factors (UV, temperature).
4. Specific characteristics of the microbial strain: microbial strains with multiple PGP traits are preferable over microbial strains characterized by only one PGP, because they can reduce different stresses simultaneously.

Future crop breeding programs should prioritize the ability of plants to form stable symbiotic relationships with beneficial microorganisms, which enhance stress resistance, productivity, and resilience. In-depth characterization of microbial functions and interactions through next-generation sequencing (NGS) and real-time monitoring is crucial for selecting specific bio-stimulants tailored to crops and conditions. Studies indicate that microbes from the host plant's microbiome are more effective than non-indigenous inoculants, making native microbiome characterization essential. The development of synthetic microbial communities offers an opportunity to improve the reliability of biostimulants, though it requires careful consideration of microbial interactions. Microbial biostimulants offer a sustainable strategy to mitigate climate change-induced stresses and maintain agro-ecosystem balance, reducing the need for pesticides and heavy metals. However, regulatory and research challenges must be addressed to enhance product efficacy and adoption.

Nanoscale Materials in Horticultural Crops' Disease Management, Preventing Disease by Enhancing Plant Health with Nano-Fertilizers

Engineered nanomaterials (ENMs) can play a crucial role in plant disease management as nano fertilizers, providing a more efficient and effective means of delivering nutrients to plants. This can be achieved by using ENMs made from micro- or macronutrients, or by using nano carriers to encapsulate them and deliver the nutrients to plants. These approaches can be used to facilitate the controlled release of nutrients to plants at the right dosage and to the desired target resulting in more efficient uptake and utilization of nutrients by the crop (Adisa *et al.* 2019). Additionally, nano-fertilizers can be designed to release nutrients at specific times or in response to specific environmental conditions, providing a more tailored and effective approach to fertilization. This can be particularly beneficial in managing plant diseases, as healthy and well-nourished plants are better equipped to resist and recover from infections (Gupta *et al.* 2022). Moreover, plants that are already weak and stressed due to malnutrition will be more likely to become infected by pathogens (Cao *et al.* 2021). Nano-fertilizers are an alternative to conventional nutrient management systems that often result in the loss or inefficiency of applied fertilizers. Also, nano fertilizers of nitrogen (N), phosphorus (P), potassium (K), copper(Cu), iron (Fe), manganese (Mn), molybdenum (Mo) and zinc (Zn),can be designed to deliver nutrients more effectively only under specific soil moisture, temperature and pH conditions (Adisa *et al.* 2019), interestingly, carbon nano tubes as well as other organic and in-organic nanomaterials have been shown to have antimicrobial properties and some of those nanomaterials are composed of essential elements for plant growth (Wang *et al.* 2014). This has led to the possibility of using these nutrients in nanoparticle (NP) form to suppress plant diseases and promote plant growth simultaneously. For example, Imada *et al.* (2016) investigated the effects of magnesium oxide nanoparticles (MgO-NPs) on inducing resistance of tomato plants to *Ralstonia solanacearum*, the causal agent of bacterial wilt, and their antibacterial activity. Roots treated with 0.7 per cent MgO-NPs suspension 7 days before being infected with the pathogen significantly reduced the incidence of disease. Elmer and White (2016) observed that application of different m*et al* oxides on tomato plants cultivated in soils infested with *Fusarium* spp. rapidly generated reactive oxygen species (ROS) in the roots and significant reduction in the area under the disease progress curve (AUDPC).

Nano-particles for Managing weeds and Pests to Prevent Plant Disease Infections

Some pathogens can infect a wide range of plant species, including both cultivated and non-cultivated plants, while others can infect both weeds and crops due to close genetic similarities between those plants. This allows the transference of inoculum within plants in a field crop and between neighbouring fields. Nightshade weeds (*Solanum sarrachoides*), which are members of the Solanaceae family, are a good example of noxious weeds that can act as hosts of pathogens. These weeds are hosts for many viruses, including potato virus Y (PVY) (Cervantes), the virus can sustain over winter in those plants and become a source of inoculum for solanaceous crops during growing seasons. Water, wind, and human activities can spread pathogens or introduce them from outside sources to cultivated fields. While insect scan exacerbate this dynamic by visiting multiple hosts, including infected weed plants, and moving pathogens from plant to plant. The complex interactions between noxious weed plants, vectors, and pathogens in agricultural fields present a challenge in the development of disease management strategies. Nanoscale technologies significantly enhance pesticide effectiveness and reduce environmental and food chain impacts. Encapsulating atrazine in polymeric nanocarriers increased its efficacy tenfold against weeds like *Bidens pilosa* and *Amaranthus viridis* compared to conventional formulations (Sousa *et al.* 2018). Furthermore, nano formulations can promote the reduction of herbicide loss and the controlled release of active ingredients (Shan *et al.* 2022). The authors demonstrated that light-triggered 2,4-D nanoparticles (NPs) offer effective herbicidal action with reduced toxicity to non-target organisms compared to conventional formulations. However, a major challenge in developing nano herbicides is optimizing targeted delivery to minimize crop damage, as weeds often share similar morphology and physiology with crops. Therefore, species-specific delivery strategies need further exploration (Santana *et al.* 2022) achieved a 70% increase in cargo delivery using carbon-based nanomaterials with peptides targeting chloroplasts. This research enhances targeted nano-formulation delivery in plants, though specificity and crop selectivity need further exploration. Thiamethoxam encapsulated in polymeric micelles effectively managed Asian citrus psyllid at half the dose of non- encapsulated thiamethoxam (Assalin *et al.* 2022). Combining bio-insecticides with nano- based materials can manage pests environmentally (Campos *et al.* 2019). Imidacloprid in mesoporous silica NPs increased insecticidal activity against *Aphis craccivora* with temperature rise (Yao *et al.* 2021). RNA interference (RNAi) with nano clay blocked bean common mosaic virus (BCMV) transmission by *Myzus persicae* in *Nicotiana benthamiana* and cowpea (Worrall *et al.* 2019).

Nanoparticle-based methods enable rapid, sensitive field diagnostics for early plant disease detection, enhancing traditional methods like PCR and antibody assays (Thangavelu *et al.* 2022). Gold nanoparticles (Au-NPs) functionalized with pathogen-specific antibodies can detect pathogens via localized surface plasmon resonance (LSPR), changing color based on Au-NP size and aggregation (Kulabhusan *et al.* 2022). This technique detects small amounts of pathogens, including viruses, fungi, and bacteria (Dharanivasan *et al.* 2016). Detection methods include SERS, fluorescence microscopy, electron microscopy, and electrochemical sensing (Martinelli *et al.* 2015). Besides Au-NPs, silver, carbon nanomaterials, quantum dots, and m*et al* oxides with probes are used in plant disease diagnosis (Rad *et al.* 2012). For fungal diagnosis, stannic oxide (SnO_2) and titanium dioxide (TiO2) NPs detect p-ethylguaiacol, a compound from Phytophthora cactorum, which infects crops like strawberries and cucurbits (Fang *et al.* 2014). Using m*et al* oxide NPs of SnO_2 and TiO2 instead of gold electrodes enhances sensitivity and reduces costs for detecting low levels of p-ethylguaiacol, enabling early detection of *P. cactorum* infections. Nanoscale materials in biosensors or wearable sensors, combined with portable devices like smartphones, allow fast, on-site monitoring of plant diseases through indicators like changes in plant morphology, growth rate, and volatile compound emissions (Lee *et al.* 2021). For example, an electrochemical sensor with a gold electrode modified with Cu-NPs can monitor salicylic acid (SA) levels to detect *Sclerotinia sclerotiorum* in oilseeds (Ariffin *et al.* 2014). NP-based methods offer high sensitivity, specificity, and simplicity, making them suitable for field use.

Role of Nano-Scale Materials in Managing fungal and Fungal-Like Pathogens

Nanotechnology is being tested as a powerful tool to aid the management of fungal diseases in diagnosis, plant nutrition, nano-carriers of fungicides, and the direct effect of NPs in managing fungal pathogens (Elmer & White 2018). For example, rice plants infected with *Fusarium fujikuroi* have impaired absorption and translocation of nutrients due to rot in the crown and abnormal stem elongation (Chen *et al.* 2020). However, adding nanoscale nutrients, such as a combination of Cu with Fe, S, Ca, Zn, and Mn improved nutrient uptake and movement in the infected plants and generated an immune response to the fungus (Shang *et al.* 2020). The authors found that Ca can act as a secondary messenger to activate the plant defense system, while Mn can enhance pathogen-induced lignification and callus formation, reducing localized cell death in the plant (Eskandari *et al.* 2020). Copper (Cu) has long been used in the field to manage fungal diseases of plants; however, there are many concerns about the high levels of Cu that can accumulate in the environment, leading to resistance among pathogens along with other environmental issues. To optimize the effectiveness of copper-based fungicides, nano materials such as copper oxide (CuO), copper sulfide (CuS), copper hydroxide ($Cu(OH)_2$, and cupric phosphate tri hydrate ($Cu_3(PO_4)_2 \cdot 3H_2O$) nano sheets were viable alternatives to traditional methods. In a breakthrough study, foliar applications of CuO and $Cu_3(PO_4)_2 \cdot 3H_2O$ NPs were able to reduce the symptoms caused by *Fusarium oxysporum* f.sp. *niveumin* watermelon by 29.9 and 39.2%, respectively, compared with a conventional product ($CuSO_4$ salt in bulk) and resulted in a 38.2% increase in fruit yield (Borgatta *et al.* 2018). Post harvest coating of fruits and vegetables with organic compounds combined with nano materials like chitosan-coated iron oxide (CH-Fe_2O_3) NPs can effectively control the growthof fungi such as *Rhizopus stolonifer* (bread mold).

Role of Nanotechnology in Managing Bacterial Diseases

Nanoparticles are being employed to defy bacterial pathogens in three ways, mainly: (i) as carriers of antibacterial chemicals, (ii) to boost plants' defenses against bacteria, and (iii) directly interactingwith bacterial cells leading to cell death. Nanoparticles as plant defense boosters NPs enhance plant defense mechanisms and protect against a range of pathogens by activating the accumulation of secondary metabolite (SMA), ROS, pathogenesis-related proteins (PRP), and defense-related enzymes (DRE), as well as augmenting hypersensitive response (HR) and complex signal transduction (CSR) (El-Shetehy *et al.* 2021). For instance, nanoparticles (NPs) can boost defense-related plant hormones like SA and activate SAR in plants, helping prevent bacterial infections and reducing damage to crops (Imada *et al.* 2016). Nano-biotics, with antibacterial properties, can control a wide range of plant pathogenic bacteria. These nanomaterials attach to bacterial cells through electrostatic attraction, leading to cell damage (Wagi & Ahmed 2020). However, bacterial sensitivity to NPs may vary due to growth rate and biofilm formation, and the exact molecular mechanisms remain unclear. One well-known mechanism is nanotoxicity, which causes oxidative stress in bacterial cells by generating ROS (Chakraborty *et al.* 2022); however, nanotoxicity can also harm plant cells (El-Shetehy *et al.* 2021). Nevertheless, a recent study reported that Cu-NPs showed antimicrobial activity against rice leaf blight disease, caused by the bacterium *Xanthomonas oryzae* pv. *oryzae,* while enhancing the growth and development of seedlings (Majumdar *et al.* 2021).

Role of Nano-Technology in Managing Viruses and Viroids

Nanoparticles such as silver (Ag), zinc oxide (ZnO), and siliconmonoxide (SiO) showed promise in reducing viral activity and infectivity in plants (Farooq *et al.* 2021). These nanomaterials can reduce virus accumulation in plant cells and activate plant defense mechanisms. When applied directly, NPs can deactivate viruses by inhibiting DNA/RNA replication, protein synthesis, capsid formation, and viral genome packaging into the capsid (Rajani *et al.* 2022). In fact, Ag-NPs applied at 50 ppm on tomato plants helped to sup-press tomato mosaic virus (ToMV) and PVY by 17.50 and 20.25%, respectively (El-Dougdoug *et al.* 2018). It binds to the viruses' coat proteins, disintegrating the virions. In another study, daily foliar spraying of ZnO-NP and SiO_2-NPs onto the leaves of tobacco for 12 days inhibited tobacco mosaic virus (TMV) replication (Cai *et al.* 2019). These properties were primarily due to the direct inactivation of TMV and improvement in plant defenses through induced resistance. A promising antiviral strategy that is being intensively investigated now is the application of exogenous double-stranded RNA (dsRNA) molecules to plants to activate RNA interference (RNAi) to suppress viral infections. In the RNAi pathway, dsRNAs are recognized and processed by Dicer-like enzymes into small interference (si) RNAs (21 to 25 nucleotides in length). Processed RNAs are incorporated into Argonaute proteins (AGOs) that guide the cleavage of target RNAs by sequence complementarity (Mello & Conte 2004). Recent studies have demonstrated the efficiency of the application of exogenous naked dsRNA for RNAi-mediated protection against a range of viruses and viroids on multiple crops (Rosa *et al.* 2018). Combining dsRNA with layered double hydroxide (LDH) significantly improves dsRNA stability under different environmental conditions (Mitter *et al.* 2017). After 30 days, applied naked dsRNAs were completely degraded, whereas the dsRNA-LDH complex was still detectable on plant leaves. This study introduced the potential of using nanocarriers to protect nucleic acids from degradation, offering sustainable crop protection against plant viruses. Since then, various studies have explored nanoparticles (NPs) as carriers of RNA molecules for plant protection (Shidore *et al.* 2021). However, no commercial RNA-NP-based product has been released for field use, highlighting the need for further research in this promising but underexplored area.

Role of Nano-Technology in Managing Nematodes

The most common way to reduce nematode populations is by applying synthetic nematicides, though methods like fumigation, solarization, organic amendments, and bio- antagonists are also used. However, synthetic nematicides are toxic and harm beneficial microbes in the rhizosphere, such as rhizobia and mycorrhizae. NPs made from silver, copper, zinc, titanium, gold, or carbon exhibit nematicidal activity when applied to nematodes at various developmental stages (Sabry 2019). This activity is measured by reductions in root galls, egg masses, juveniles, and overall nematode populations. Field experiments show NPs effectively manage nematodes, improving plant growth. NPs kill nematodes by disrupting cellular processes, altering membrane permeability, and inducing oxidative stress, similar to other antimicrobial agents. While *in-vitro* studies show promise, more comprehensive field experiments are needed to target pathogenic nematodes. Nanocarriers have been used to deliver active agents specifically against plant-parasitic nematodes. In two studies, laboratory-synthesized plant virus NPs (red clover necrotic mosaic virus and tobacco mild green mosaic virus) were used as nano carriers to deliver abamectin (abio-antagonist) and crystal violet (an anthelmintic drug) for treating pathogenic nematodes (Cao *et al.* 2015). Similarly, several nanocarriers (chitosan, zein, and microspheres) loaded with nematicide compounds have been successfully tested for their control and targeted delivery in plant and root systems (Demchak & Dybas 1997). Combining biologically derived active ingredients (such as abamectin/azadirachtin) or RNAi-inducers with nanocarrier delivery systems could be a promising nano-formulation for targeting plant parasitic nematodes.

Biosynthesized Nano-Particles and their Advantages in Plant Disease Management

NP biosynthesis is a complex multistep synthesis approach regulated by catalytic enzymes, in which natural bio-sources are used to produce the material of interest. NP biosynthesis utilizes macro- and microscopic organisms (such as plants, bacteria, fungi, yeast, and microalgae) as starting materials in the process of producing NPs. This process takes advantage of the various natural bio-reducing agents (biomolecules and secondary metabolites such as polyphenols, flavonoids, alkaloids, terpenoids, sugars, and amino acids) produced by those organisms (Naikoo *et al.* 2021). The functional groups of the bio- molecules release electrons in the reaction medium that interact with the metal salt, reducing metal ions into m*et al.*

Newly Developed Plant Disease Detection Techniques

These techniques identify plant disease symptoms like necrosis, leaf colour changes, and chlorosis. Field inspection confirms the presence of the target pathogen and allows monitoring of symptom-free plants for timely intervention. Hyperspectral imaging, combining imaging with spectroscopy, is effective but limited in field use due to high costs, long data acquisition times, and complex data interpretation. It successfully detected target spots caused by *Corynespora cassicola* and bacterial spots by *Xanthomonas perforans*, achieving 99% accuracy using a multilayer perceptron neural network under both UAV- based field and laboratory conditions.

Raman spectroscopy (RS) enables early, accurate disease detection by collecting inelastically dispersed photons, providing quick results without sample preparation. RS achieved 94% accuracy and was used for early diagnosis of rice blast disease, maize rot disease caused by *Colletotrichum graminicola*, rose rosette infection, and to investigate wheat and sorghum grains infected with ergot, black tip, or mold. RS combined with deep learning networks was recently used to detect Fusarium head blight (FHB)-infected wheat kernels.

A smartphone-based instrument for plant disease diagnosis distinguishes uninfected from infected leaf samples in 15 minutes by detecting and classifying 10 plant VOCs at parts- per-million concentrations. However, uneven pathogen distribution can cause false negatives. A piezoelectric cantilever resonator on a stretchable substrate monitors plant responses for early disease diagnosis and stress identification, achieving 89% accuracy in detecting VOC biomarkers from healthy and infected plants with late blight caused by *Phytophthora infestans.* M in ION nanopore technology, developed by Oxford Nanopore Technologies, offers instant access to gigabases of long-read data. This portable, USB-powered instrument is smaller than a smartphone and provides output data within minutes. It has been used for sequencing bacterial and virus pathogens, detecting and genotyping potato viruses, identifying *Xylella fastidiosa* subspecies in olive and grapevine samples, and detecting *Fusarium oxysporum* on *Zinnia hybrida* (zinnia) plants.

MALDI-TOF MS (Matrix-assisted laser desorption ionization time-of-flight mass spectrometry) is a sophisticated technology revolutionizing biological research, clinical settings, and microbiological diagnostics by identifying large biological molecules like proteins. Recently introduced in microbiology labs, it offers rapid, accurate, and cost- effective microorganism identification. It has been used to detect pathogenesis-related proteins in tomato plants infected by *Fusarium oxysporum* f. sp. *Radices- lycopersici* and *Ralstonia solanacearum*, and to identify *Pantoea stewartii* subsp. *stewartii*, causing Stewart's wilt in sweet corn. It also detects proteins contributing to rice resistance against *Xanthomonas oryzae* pv. *oryzicola* MALDI-TOF MS rapidly identifies different *Puccinia* species, causing rust in wheat and other crops, and detects proteins associated with Flavescence dorée in grapevine. It confirms antibacterial compounds against *Erwinia amylovora* (fire blight) and *E. carotovora* (soft rot). Additionally, it helps understand the role of *Enterobacter cloacae* in banana resistance to *Pseudocercospora fijiensis* (Black Sigatoka). However, its use in detecting viral plant pathogens is still limited.

Role of Nutrients in Controlling Plant Diseases in Sustainable Agriculture

When a plant is infected by a pathogen, its physiology, especially nutrient uptake, assimilation, translocation, and utilization, is impaired (Marschner 1995). Some pathogens immobilize nutrients in the rhizosphere or infected tissues, while others interfere with nutrient translocation or utilization, causing deficiencies or toxicities (Huber & Graham, 1999). Pathogens may also consume nutrients for their growth, reducing availability for the plant and increasing susceptibility due to deficiency (Timonin 1965). Root infections, common with soil-borne pathogens, hinder water and nutrient uptake, particularly when nutrient levels are marginal. Stem girdling or vascular infections can further limit root function and nutrient translocation. Pathogens also affect membrane permeability, leading to deficiencies or toxicities. For example, *Fusarium oxysporum,* F. *vasinfectum* increases phosphorus levels in leaves but reduces nitrogen, potassium, calcium, and magnesium (Huber & Graham 1999).

Nutrient management can influence disease severity. Fertilization improves plant growth and reduces disease under nutrient deficiency. Nitrogen application reduced take-all in cereal crops (*Gaeumannomyces graminis*), while phosphorus decreased take-all and pythium root rot (Kiraly 1976). However, increased nitrogen raised the incidence of foliar diseases like rust and powdery mildew. The nutrient-disease interaction is complex, and the impact of each nutrient on specific diseases and mechanisms of tolerance or resistance must be examined individually.

Nitrogen

Nitrogen (N) is essential for plant growth, and its impact on disease resistance is well- documented (Marschner 1995). However, studies on N's effect on disease development are inconsistent, possibly due to differences in N forms, pathogen types (obligate vs. facultative), or timing of N application (Hoffland *et al.* 2000). There is a lack of systematic studies on how N supply influences disease resistance and interactions with biocontrol agents (Tziros *et al.* 2006). The effect of N varies by pathogen type. High N levels increase infection severity from obligate parasites (e.g., *Puccinia graminis*), while reducing infection from facultative parasites (e.g., *Alternaria, Fusarium*). Soil-borne pathogens complicate matters due to competition and chemical barriers at the root surface (Huber 1980). The nutritional needs of pathogens explain these differences: obligate parasites depend on living cells, while facultative parasites target senescing tissue or release toxins (Agrios 2005).

High N levels in plants enhance growth but make young tissues, which are more vulnerable, more prominent. Elevated N also reduces defense compounds like phenolics and lignin, increasing susceptibility (Robinson & Hodges 1981). Additionally, high N decreases silicon content, weakening plant defenses (Grosse-Brauckmann 1957). Plants with low N availability show stronger pathogen defenses due to increased synthesis of defense compounds (*Bryant et al.* 1983). For obligate pathogens (e.g., *Pseudomonas syringae*), high N increases susceptibility, while for facultative pathogens (e.g., *Botrytis cinerea*), high N enhances resistance (Hoffland *et al.* 2000). The form of N also matters: high nitrate (NO3) levels reduce disease from pathogens like *Fusarium*, while high ammonium (NH4) reduces infections from others like *Pyricularia*. N levels affect soil pH, nutrient availability, and phenolic content, all influencing disease resistance. Further research is needed to clarify these complex interactions.

Phosphorus

Phosphorus (P), the second most commonly applied nutrient, is vital for plant metabolic processes, being a component of DNA, RNA, ATP, and phospholipids. Its role in disease resistance, however, is inconsistent (Kiraly1976). P is particularly beneficial in controlling seedling and fungal diseases by promoting vigorous root development, helping plants escape disease (Huber & Graham 1999). For example, phosphate fertilization nearly eliminates *pythium* root rot in wheat and reduces root rot and soil smut in corn grown on P- deficient soils (Huber1980). P application has also been shown to reduce bacterial leaf blight in rice, downy mildew, tobacco leaf curl virus, soybean pod and stem blight, barley yellow dwarf virus, sugarcane brown stripe, and rice blast (Reuveni *et al.* 2000). However, in some cases, P can increase disease severity, such as with *Sclerotinia* in garden plants, *Bremia* in lettuce, and flag smut in wheat. Foliar application of P can provide local and systemic protection against powdery mildew in crops like cucumber, roses, grapes, mango, and nectarines (Reuveni & Reuveni 1998).

Potassium

Potassium (K) decreases plant susceptibility to disease up to the optimal level for growth, but beyond this, increasing K supply does not further enhance resistance (Huber & Graham 1999). K deficiency leads to metabolic issues, impairing the synthesis of high- molecular-weight compounds like proteins and starch, while promoting the accumulation of simple organic compounds that pathogens can exploit. K also aids in forming thicker epidermal cell walls, which helps prevent disease. Deficient plants also show altered protein synthesis and accumulate nitrogen compounds that pathogens use, while tissue hardening and stomatal patterns influence infestation (Marschner 1995). K application has been shown to reduce disease intensity in both obligate and facultative parasites. It decreases diseases like *Hemlinthosporium* leaf blight in wheat (Sharma *et al.* 2005) and bacterial leaf blight, sheath blight, stem rot in rice, and black rust in wheat. Other diseases reduced by K include sugary disease in sorghum, *Cercospora* leaf spot in cassava and mungbean, and seedling rot caused by *Rhizoctonia solani* (Sharma & Duveiller 2004). The balance between N and K also affects disease susceptibility.

Calcium

Calcium (Ca) affects disease susceptibility in two key ways. First, Ca stabilizes plant membranes; when deficient, membrane leakage of sugars and amino acids into the apoplast stimulates pathogen infection (Marschner 1995). Second, Ca is vital for cell wall stability, as calcium polygalacturonates in the middle lamella maintain cell wall integrity. Low Ca levels increase susceptibility to fungi that invade the xylem, leading to wilting. Ca-deficient tissues are also more prone to parasitic diseases during storage, making pre-storage Ca treatments effective against physiological disorders and fruit rot. Adequate soil Ca protects peanut pods from *Rhizoctonia* and *Pythium* infections, and soil Ca applications can prevent disease (Huber 1980). Ca also confers resistance against *Pythium*, *Sclerotinia*, *Botrytis*, and *Fusarium* (Graham 1983). In alfalfa, Ca supports the growth of *Colletotrichum trifolii* by enhancing the pathogen's pectolytic enzyme activity (Kiraly 1976). Ca may protect against *Sclerotinia sclerotiorum* by binding oxalic acid or strengthening cell walls.

Other Nutrients

Regarding other nutrients such as sulphur and magnesium, there is not enough information about their role in plant diseases. Sulphur can reduce the severity of potato scab, whereas Mg decreases the Ca content of peanut pods and may predispose them to pod breakdown by *Rhizoctonia* and *Pythium* (Huber 1980).

Micronutrients

Micronutrients reduce disease severity by influencing plant physiology and biochemistry, impacting the plant's response to pathogens (Marschner 1995). Deficient plants, with weakened defenses, may also become more susceptible to feeding as metabolites like sugars and amino acids leak from cells. For example, Zn-deficient plants showed increased disease severity from *Oidium* spp., and B-deficient wheat exhibited faster fungal spread (Bolle-Jones & Hilton 1956). Micronutrients like Mn, Cu, and B may trigger systemic acquired resistance (SAR) by releasing Ca^{2+} ions from cell walls, which interact with salicylic acid (Reuveni *et al.* 1997). Additionally, micronutrients enhance plant metabolism by influencing phenolic content, lignin, and membrane stability (Graham & Webb 1991).

Manganese (Mn) is a key micronutrient in plant resistance to root and foliar diseases (Graham & Webb 1991). Mn requirements are higher in plants than in fungi and bacteria, allowing pathogens to exploit this difference (Marschner 1995). Mn fertilization helps control diseases like powdery mildew, downy mildew, and others (Brennan 1992), though its use is limited due to its poor residual effect and rapid immobilization in soils like calcareous soils. Mn plays a crucial role in lignin and phenol biosynthesis, photosynthesis, and inhibiting fungal enzymes that degrade host cells (Römheld & Marschner 1991). It supports the production of lignin and suberin, which form biochemical barriers to fungal invasion (Hammerschmidt & Nicholson 2000). Mn has been shown to reduce diseases such as potato scab, *Fusarium* infections in cotton, and *Sclerotinia* in squash (Agrios 2005).

Zinc (Zn) has varied effects on plant disease susceptibility, sometimes reducing, increasing, or having no effect. In most cases, Zn application reduces disease severity, likely due to its direct toxic effects on pathogens rather than through plant metabolism (Graham & Webb 1991). Zn is vital for protein and starch synthesis, and its deficiency leads to the accumulation of amino acids and sugars in plant tissues (Marschner 1995). As an activator of Cu/Zn-SOD, Zn protects membranes from oxidative damage by detoxifying superoxide radicals (Cakmak 2000). Zn application has been shown to reduce *Fusarium graminearum* and root rot infections in wheat (Graham & Webb 1991).

Boron (B) is an essential yet poorly understood micronutrient, though B deficiency is the most widespread globally (Brown *et al.* 2002). It plays a key role in cell wall structure and stability, potentially reducing disease severity. Its effects on disease resistance may relate to its roles in cell wall structure, membrane permeability and stability, or phenolic and lignin metabolism (Blevins & Lukaszewski 1998). It has been shown to reduce diseases such as clubroot in crucifers, *Fusarium solani* in beans, *Verticillium* in tomatoes, and wheat powdery mildew (Marschner 1995).

Iron (Fe) is a crucial micronutrient for animals and humans but its role in plant disease resistance is less understood. Unlike Mn, Cu, and B, pathogens like Fusarium often have higher Fe requirements than plants, complicating

Fe's effects on disease resistance. While Fe can reduce diseases like wheat rust and banana anthracnose, its application can also have mixed outcomes, such as increasing or not affecting disease severity (Graham & Webb 1991). In cabbage, Fe overcame fungus-induced deficiency but didn't alter infection levels. Fe's variable effects stem from its role in activating enzymes involved in both pathogen infection and plant defense. Additionally, soil bacteria produce siderophores that lower Fe levels and inhibit certain pathogens, such as *Fusarium oxysporum* (Romheld & Marschner 1991).

Chlorine is required in very small amounts for plant growth and Cl deficiency has rarely been reported as a problem in agriculture. However, there are reports showing that Cl application can enhance host plants' resistance to disease in which fairly large amounts of Cl are required, which are much higher than those required to fulfill its role as a micronutrient but far less than those required to induce toxicity (Mann *et al.* 2004). It has also been suggested that Cl might interact with other nutrients such as Mn. Cl has been shown to control a number of diseases such as stalk rot in corn, stripe rust in wheat, take all in wheat, northern corn leaf blight and downy mildew of millet, and septoria in wheat (Graham & Webb 1991; Mann *et al.* 2004). The mechanism of Cl's effect on resistance is not well understood. It appears to be non-toxic in vitro and does not stimulate lignin synthesis in wounded wheat leaves. It was suggested that Cl can compete with NO^{-3} absorption and influences the rhizosphere pH: it can suppress nitrification and increase the availability of Mn. Furthermore, Cl ions can mediate reduce the Mn III, IV oxides and increase Mn for the plant, increasing the tolerance to pathogens.

Chlorine (Cl) is required in very small amounts for plant growth, with Cl deficiency rarely reported in agriculture. However, Cl application can enhance host plants' resistance to disease, requiring larger amounts than as a micronutrient but less than toxic levels (Mann *et al.* 2004). Cl might interact with other nutrients like Mn and control diseases such as stalk rot in corn, stripe rust in wheat, and downy mildew of millet (Graham & Webb 1991; Mann *et al.* 2004). The mechanism of Cl's effect on resistance is not well understood. It may compete with NO^{-3} absorption, influence rhizosphere pH, suppress nitrification, and increase Mn availability, boosting pathogen tolerance.

Silicon (Si) is the second most abundant element in soil and a component of plants, though not essential except for the Equisitaceae family (Marschner 1995). Adding Si to soil improves growth, yield, reduces mineral toxicities, and enhances disease and insect resistance in plants low in soluble Si (Alvarez and Datnoff 2001). Crops like rice and sugarcane, which accumulate high Si levels, are often fertilized with calcium silicate slag for better yields and disease resistance. Si controls diseases such as blast in St. Augustine grass, brown spot in rice, and sheath blight in rice, and increases turfgrass tolerance to various pathogens (Alvarez & Datnoff 2001). The mechanism is not well understood but may involve creating a physical barrier or inducing antifungal compounds like flavonoids and diterpenoid phytoalexins (Alvarez & Datnoff 2001).

Besides essential nutrients, several trace elements (Li, Na, Be, Al, Ge, F, Br, I, Co, Cr, Cd, Pd, and Hg) can be found in plant tissue and are sometimes linked to host-pathogen relationships. Notably, Li and Cd suppress powdery mildews. Cd inhibits spore germination and development at 3 mg kg^{-1}, a non-toxic level that triggers a host response. Cd and Hg also promote lignin synthesis in wheat (Graham and Webb 1991). The mechanism of Li is unclear, but it may catalyze a defense-related metabolic pathway.

Use of Cultural Methods in Improving Plant Nutrition and Disease Resistance

Not only can nutrient application as fertilizers increase disease tolerance, but measures that enhance nutrient availability and balance can also affect plant growth and disease resistance. Sustainable agriculture approaches like crop rotation, green manure, manure application, intercropping, and tillage improve plant nutrition and disease tolerance (Oborn *et al.* 2003). These methods increase soil organic matter (SOM), crucial for soil health and plant growth. Organic residues (crop residues, cover crops, organic wastes) affect soilborne pathogens and nutrient availability (Stone *et al.* 2004). Adding sphagnum peat, green manures, and animal manures can create suppressive soils, reducing pathogen establishment (Hu *et al.* 1997). Dairy manure suppresses pathogens in sweet corn (*Drechslera* spp., *Phoma* spp., *Pythium arrhenomanes*) and snap bean (*Fusarium solani, Pythium* spp.). Mechanisms for disease suppression include microbiostasis, pathogen propagule destruction, antibiosis, substrate competition, and induced systemic resistance (SAR). SOM impacts nutrient content and availability through soil microorganisms, affecting disease incidence by enhancing plant resistance and growth. Studies show fields with organic amendments have higher microbial activity and nutrient content, reducing disease incidence (Loschinkohl & Boehm 2001).

Crop Rotation and Cover Crops

Crop rotation involves growing different crops sequentially on the same field. Long- term studies show that crop rotation, combined with fertility management, is crucial for sustainable agriculture (Reid *et al.* 2001). It controls diseases by starving plant pathogens in the soil through non-host crops (Reid *et al.*

2001). In beans, rotation is highly effective for disease control. It can increase nitrogen (N) levels and affect other nutrients, influencing disease severity (Reid *et al.* 2001). For example, rotating with lupins increases manganese (Mn) availability (Graham and Webb 1991).

Cover crops also impact soil properties and microbial communities, affecting disease severity. They increase soil organic matter (OM), microbial biomass, and activity, contributing to disease suppression. Crop rotation enhances soil buffering capacity, denies pathogens a host, and affects nitrification, influencing soil N forms (Graham and Webb 1991). Green manures improve N and other nutrients like phosphorus (P) and potassium (K), significantly impacting disease development. They can fix N with bacteria, increasing soil N levels by 459 Kg N/ha (Cherr *et al.* 2006), and affect nutrients like P, Mn, and zinc (Zn), enhancing disease tolerance.

Intercropping

Intercropping systems have the potential to reduce the incidence of diseases. However, different responses to disease severity with different systems of intercropping have been observed. There are four mechanisms involved in an intercropping system that can reduce disease incidence, all of which involve lowering the population growth rate of the attacking organism:

1. The associate crop causes plants of the attacked component to be poorer hosts.
2. The associate crop interferes directly with the attacking organism.
3. The associate crop changes the environment of the host such that natural enemies of the attacking organism are favored.
4. The presence of non-host or resistant plants growing in between susceptible plants can physically block inoculum from reaching the susceptible hosts (*i.e.* the non-host serves as a physical barrier to the pathogen inoculum). Francis (1989) found that intercropping reduced pests and diseases in 53% of experiments but increased them in 18%. Reasons for increased pests include reduced cultivation, increased shading, alternative hosts, and crop residues serving as pathogen inoculums. Intercropping improves nutrients by increasing nitrogen (N) from legumes and enhancing phosphorus (P) and potassium (K) uptake. Reduced tillage systems, including zero tillage, increase soil organic matter (SOM), conserve SOM, reduce erosion, and lower energy consumption and production costs (Carter 1994). However, reduced tillage can alter the soil environment, affecting disease incidence variably.

Minimum tillage concentrates residues and pathogen propagules on the soil surface, potentially impacting disease incidence. Conservation tillage has increased stubble-borne diseases due to slower colonization by soil organisms and favored pathogen survival (Bockus & Shroyer 1998). Conservation tillage systems also concentrate plant residues in the surface soil layer, increasing microbial biomass and activity (Dick 1984).

Systemic Induced Resistance or Systemic Acquired Resistance

The induction of resistance reactions in plants against pathogens is a well-known phenomenon in plant pathology. Initially described as resistance to nonvirulent pathogens, this enduring, non-specific resistance is called systemic acquired resistance (SAR) when systemically distributed within the plant. SAR can be induced by avirulent pathogens or chemical compounds like salicylic acid (SA), which is involved in the signal transduction pathway leading to SAR. Structural analogues of SA can also induce SAR. Wiese *et al.* (2003) introduced the term chemically-induced resistance (CIR) to describe systemic resistance after applying synthetic compounds. This resistance involves forming structural barriers such as lignification and induction of pathogen-related proteins (Graham & Webb 1991). Systemic induced resistance (SIR) can be induced by foliar sprays of nutrients like phosphates, potassium (K), and nitrogen (N). It is hypothesized that during SIR, an immunity signal released or synthesized at the induction site is systemically translocated to the challenged leaves, activating defense mechanisms. Although SA is hypothesized as a possible signal, its presence in phloem sap of non-infected upper leaves but not in infected lower leaves suggests it may not be the primary systemic signal for SIR (Reuveni & Reuveni 1998).

A single phosphate foliar application can induce high levels of systemic protection against powdery mildew in cucumbers (*Sphaerotheca fuliginea*) (*Reuveni et al.* 1997a, b). Similar responses were found in maize, where foliar spray with phosphates induced systemic protection against common rust (*Puccinia sorgi*) and northern leaf blight (NLB) (*Exserohilum turcicum*). Trace elements also play a crucial role in plant disease resistance. Foliar sprays with H_3BO_3, $CuSO_4$, $MnCl_2$, or $KMnO_4$ separately induced systemic protection against powdery mildew in cucumbers. In wheat, applications of boron (B), manganese (Mn), and zinc (Zn) increased resistance to tan spot (Simoglou & Dordas 2006).

The mechanism of SIR development is still unknown. It is proposed that chemicals trigger the release and rapid movement of the "immunity signal" from infected to unchallenged leaves (Reuveni & Reuveni 1998). This mechanism might involve increased peroxidase activity and β-1,3-glucanase in protected leaves. Mn and Cu might act as cofactors of met all oprotein enzymes like peroxidase, which Mn ions induce (Mengel & Kirkby 2001). Peroxidase and β-1,3-glucanase are involved in cross-linking cell wall components, polymerizing lignin and suberin monomers, and subsequent pathogen resistance.

SA is proposed as a translocatable signal compound in SIR, interacting with intercellular Ca^{2+} in chitinase induction in carrot suspension culture. Applications of Mn, Cu, and B can increase Ca^{2+} cations, interact with SA, and activate SIR (Reuveni & Reuveni 1998). These findings indicate that resistance mechanisms are present in susceptible plants and can be induced by simple inorganic chemicals, and this induced resistance is not pest-specific.

Big Data and Machine Learning Approaches in Crop Protection

This section provides an overview of technologies and potential applications in crop protection using Big Data and machine learning. It covers four main applications: (i) Prediction and modeling of herbicide resistance; (ii) Detection and management of invasive species and weeds; (iii) Decision support systems for crop protection; and (iv) Robotics and autonomous weed control systems. Key components of Big Data, such as data acquisition, storage, and analytics, are briefly discussed, followed by a review of popular machine learning techniques, including discriminative/generative and supervised/unsupervised learning approaches. A recent review by Heap (2014) highlighted the increasing herbicide resistance in agricultural weeds due to persistent herbicide use. In Australia, significant research focuses on quantifying herbicide resistance, particularly in annual ryegrass (*Lolium rigidum*) (Boutsalis *et al.* 2012). Intelligent-based approaches have been proposed for herbicide modeling and prediction. Diaz *et al.* (2005) used a rule-based machine learning model to predict wild-oat (*Avena sterilis* L.) density, explaining about 88% of weed variability. Evans *et al.* (2016) modelled glyphosate resistance in *Amaranthus tuberculatus* using classification and regression trees (CART), showing that herbicide mixing reduces glyphosate resistance. Machine learning is also applied to detect invasive species and weeds.

Lawrence *et al.* (2006) used random forest classifiers to detect invasive plants (leafy spurge and spotted knapweed) from aerial hyperspectral imagery, achieving 84% and 86% accuracy, respectively. Schmidt and Drake (2011) used boosted regression trees to study why some plant genera are more invasive, explaining 24% and 29% of the variation in invasiveness. Alexandridis *et al.* (2017) used four novelty detection classifiers (OC-SVM, OC-SOM, Autoencoders, OC-PCA) to identify *S. marianum* from multispectral UAV imagery, with accuracy rates of 90-96%. Machine learning and Big Data are also applied in decision support systems (DSS) for crop protection. Early DSS models used simpler techniques like regression (Knight 1997). Modern DSS, such as Nemadecide (Been *et al.* 2005), RIM (Lacoste and Powles 2015), Blight Pro (Small *et al.* 2015), and CPO-Weeds (Sønderskov *et al.* 2016), employ advanced machine learning and Big Data techniques for more sophisticated crop protection features. Nemadecide supports strategic decisions for managing potato cyst nematodes using mathematical models developed over fifty years.

The models consider plant growth, tolerance, population dynamics, plant resistance, nematode virulence, spatial distribution, and sampling methods. RIM (Ryegrass Integrated Management) is a DSS for evaluating strategies to control ryegrass (*Lolium rigidum*) using population dynamic and rule-based models. It includes aspects of the ryegrass lifecycle like germination, survival, competition, seed production, and seed bank persistence. Blight Pro is a DSS for managing potato and tomato late blight, predicting disease dynamics based on weather, crop information, and management practices. It uses Blitecast to predict initial blight occurrence and Simcast to account for host resistance and fungicide weathering. CPO-Weeds is a knowledge-driven DSS for weed control in Denmark, providing herbicide dose recommendations based on a large database of herbicide efficacies. The review highlights the growing role of robotics and autonomous weed control systems. Examples include the RIPPA (Robot for Intelligent Perception and Precision Application), an autonomous vehicle for crop mapping and weed detection developed at Sydney University.

Future Perspectives

Research is needed to identify nutrients or combinations that reduce disease severity and to develop integrated pest management (IPM) strategies using disease-resistant varieties and cultural techniques. Understanding how nutrients affect disease tolerance, plant metabolism, and biochemical pathways is crucial. Mild nutrient deficiencies often impact secondary metabolism, which is linked to pathogen defense. Recognizing elements essential for defense mechanisms may require revising essentiality criteria. Systemic induced resistance (SIR) through nutrient application could reduce disease severity. Products like Actigard activate similar defense responses. Combining NPK fertilizers with disease-resistant cultivars and other nutrients, along with crop rotation and organic amendments, can enhance nutrient availability and reduce disease in sustainable agriculture.

References

Adisa, I.O., Pullagurala, V.L.R., Peralta-Videa, J.R., Dimkpa, C.O., Elmer, W.H., Gardea-Torresdey, J.L. and White, J.C. 2019. Recent advances in nano-enabled fertilizers and pesticides: A critical review of mechanisms of action. Enviromental Science Nano 6:2002-2030.

Agrios N.G. (2005) Plant Pathology, 5th ed., Elsevier-Academic Press, p. 635.

Alexandridis, T. K., Tamouridou, A. A., Pantazi, X. E., Lagopodi, A. L., Kashefi, J., Ovakoglou, G. & Moshou, D. 2017. Novelty detection classifiers in weed mapping: Silybum marianum detection on UAV multispectral images. Sensors 17(9):

Alvarez J., Datnoff L.E. 2001 The economic potential of silicon for integrated management and sustainable rice production, Crop Prot. 20, 43–48.

Ariffin, S. A. B., Adam, T., Hashim, U., Faridah Sfaridah, S., Zamri, I., and Uda, M. N. A. 2014. Plant diseases detection using nanowire as biosensor transducer. Adv. Mater. Res. 832:113-117.

Assalin R. M., de Souza, D. R. C., Rosa, M. A., Duarte, R. R. M., Castanha, R. F., Vilela, E. S. D., and Durán, N. 2022. Thiamethoxam used as nanopesticide for the effective management of Diaphorinacitri psyllid: an environmental-friendly formulation. International Journal of Pest Management, 1–9. https://doi.org/10.1080/09670874.2022.2042425

Been, T. H., Schomaker, C. H., & Molendijk, L. P. G. 2005. NemaDecide: a decision support system for the management of potato cyst nematodes. In Potato in progress (pp. 143-155). Wageningen Academic.

Blevins D.G. and Lukaszewski K.M. 1998 Boron in plant structure and function. Annu. Rev. Plant Phys. 49: 481–500.

Bockus W.W., Schroyer J.P. 1998 The impact of reduced tillage on soilborne plant pathogens, Annu. Rev. Phytopathol. 36, 485–500.

Boutsalis, P., Gill, G. S., & Preston, C. 2012. Incidence of herbicide resistance in rigid ryegrass (Lolium rigidum) across southeastern Australia. Weed Technology, 26(3), 391-398.

Brennan R.F. 1992. The role of manganese and nitrogen nutrition in the susceptibility of wheat plants to take-all in western Australia, Fertilizer Res. 31, 35–41

Brown PH, Bellaloui N, Wimmer MA, Bassil ES, Ruiz J, Hu H, Pfeffer H, Dannel F, Römheld V. 2002. Boron in plant biology. Plant Biology 4: 205–223.

Bryant J.P., Chapin F.S., Klein D.R. 1983. Carbon/nutrient balance of boreal plants in relation to vertebrate herbivory, Oikos 40, 357–368.

Cai, L., Liu, C., Fan, G., Liu, C., and Sun, X. 2019. Preventing viral disease by ZnONPs through directly deactivating TMV and activating plant immunity in Nicotiana benthamiana. Environ. Sci. Nano 6:3653-3669.

Cakmak I.M. 2000. Possible roles of zinc in protecting plant cells from damage by reactive oxygen species, New Phytol. 146, 185–205

Campos, E. V. R., Proença, P. L. F., Oliveira, J. L., Bakshi, M., Abhilash, P. C., and Fraceto, L. F. 2019. Use of botanical insecticides for sustainable agriculture: Future perspectives. Ecol. Indic. 105:483-495.

Cao, J., Guenther, R. H., Sit, T. L., Lommel, S. A., Opperman, C. H., and Willoughby, J. A. 2015. Development of abamectin loaded plant virus nanoparticles for efficacious plant parasitic nematode control. ACS Appl. Mater. Interfaces 7:9546-9553.

Cao, X., Wang, C., Luo, X., Yue, L., White, J. C., Elmer, W., Dhankher, O. P., Wang, Z., and Xing, B. 2021. Elemental sulfur nanoparticles enhance disease resistance in tomatoes. ACS Nano 15:11817-11827.

Carter M.R. 1994. Strategies to overcome impediments to adoption of conservation tillage, in: Carter M.R. (Ed.), Conservation tillage in temperate agroecosystems, Lewis Publishers, Boca Raton, FL, pp. 1–22.

Chakraborty, N., Jha, D., Roy, I., Kumar, P., Gaurav, S. S., Marimuthu, K., Ng, O.-T., Lakshminarayanan, R., Verma, N. K., and Gautam, H. K. 2022. Nanobiotics against antimicrobial resistance: Harnessing the power of nanoscale materials and technologies. J. Nanobiotechnol. 20:375.

Chen, C.-Y., Chen, S.-Y., Liu, C.-W., Wu, D.-H., Kuo, C.-C., Lin, C.-C., Chou, H.-P., Wang, Y.-Y., Tsai, Y.-C., Lai, M.-H., and Chung, C.-L. 2020. Invasion and colonization pattern of Fusarium fujikuroi in rice. Phytopathology 110:1934-1945.

Cherr C.M., Scholberg J.M.S., McSorley R. 2006. Green manure approaches to crop production: a synthesis, Agron. J. 98, 302–319.

Datta Majumdar, T., Ghosh, C. K., and Mukherjee, A. 2021. Dual role of copper nanoparticles in bacterial leaf blight-infected rice: A therapeutic and metabolic approach. ACS Agric. Sci. Technol. 1:160-172.

Demchak, R.J. and Dybas, R.A. 1997. Photostability of abamectin/zein microspheres. J. Agric. Food Chem. 45:260-262.

Dharanivasan, G., Mohammed Riyaz, S. U., Michael Immanuel Jesse, D., Raja Muthuramalingam, T., Rajendran, G., and Kathiravan, K. 2016. DNA templated self- assembly of gold nanoparticle clusters in the colorimetric detection of plant viral DNA using a gold nanoparticle conjugated bifunctional oligonucleotide probe. RSC Adv. 6:11773-11785.

Diaz, B., Ribeiro, A., Ruiz, D., Barroso, J. and Fernandez-Quintanilla, C. 2005. A genetic algorithm approach to discover complex associations between wild-oat density and soil properties. In: Proceedings of the Fourth European Conference on Precision Agriculture, edited by J. Stafford and A. Werner (Wageningen Academic publishers, The Netherlands) pp. 149–157.

Dick W.A. 1984. Influence of long-term tillage and rotation combinations on soil enzyme activities, Soil Sci. Soc. Am. J. 48, 569–574. Disease, in: Rengel Z. (Ed.), Mineral nutrition of crops fundamental mechanisms and early detection of sugarcane mosaic viruses. Sci. Rep. 12:4144.

El-Dougdoug, N. K., Bondok, A. M., and El-Dougdoug, K. A. 2018. Evaluation of silver nanoparticles as antiviral agent against ToMV and PVY in tomato plants. Middle East J. Appl. Sci. 8:100-111.

Elmer, W. H., and White, J. C. 2016. The use of m*et al*lic oxide nanoparticles to enhance growth of tomatoes and eggplants in disease infested soil or soilless medium. Environ. Sci. Nano 3:1072-1079.

Elmer, W., and White, J. C. 2018. The future of nanotechnology in plant pathology. Annu. Rev. Phytopathol. 56:111-133.

El-Shetehy, M., Moradi, A., Maceroni, M., Reinhardt, D., Petri-Fink, A., RothenRutishauser, B., Mauch, F., and Schwab, F. 2021. Silica nanoparticles enhance disease resistance in Arabidopsis plants. Nat. Nanotechnol. 16:344-353.

Eskandari, S., Höfte, H., and Zhang, T. 2020. Foliar manganese spray induces the resistance of cucumber to Colletotrichum lagenarium. J. Plant Physiol. 246- 247:153129.

Evans, J.A., Tranel, P.J., Hager, A.G., Schutte, B., Wu, C., Chatham, L.A., Davis, A.S. 2016.
Managing the evolution of herbicide resistance. Pest Management Science 72(1): 74-80.

Fang, Y., Umasankar, Y., and Ramasamy, R. P. 2014. Electrochemical detection of p-ethylguaiacol, a fungi infected fruit volatile using m*et al* oxide nanoparticles. Analyst 139:3804-3810.

Graham D.R., Webb M.J. 1991. Micronutrients and disease resistance and tolerance in plants, in: Mortvedt J.J., Cox F.R., Shuman L.M., Welch R.M. (Eds.), Micronutrients in Agriculture, 2nd ed., Soil Science Society of America, Inc. Madison, Wisconsin, USA, pp. 329–370.

Grosse-Brauckmann E. 1957. The influence of silicic acid on mildew infection of cereals with different nitrogen fertilizers, Phytopathol. Z. 30, 112–115.

Gupta, R., Leibman-Markus, M., Anand, G., Rav-David, D., Yermiyahu, U., Elad, Y., and Bar, M. 2022. Nutrient elements promote disease resistance in tomato by differentially activating immune pathways. Phytopathology 112:2360-2371.

Hammerschmidt R., Nicholson R.L. 2000. A survey of plant defense responses to pathogens, in: Agrawal A.A., Tuzun S., Bent E. (Eds.), Induced plant defenses against pathogens and herbivores, APS Press, Minneapolis, USA, p. 390.

Hoffland E., Jegger M.J., van Beusichem M.L. 2000. Effect of nitrogen supply rate on disease resistance in tomato depends on the pathogen, Plant Soil 218, 239–247.

Hu S., van Bruggen A.H.C., Wakeman R.J., Grunwald N.J. 1997. Microbial suppression of in vitro growth of Pythium ultimum and disease incidence in relation to soil C and N availability, Plant Soil 195, 43–52.

Huber D.M. 1980. The role of mineral nutrition in defense. In Plant Disease, An advancedimplications, Food Product Press, New York, pp. 205–226.

Huber D.M., Graham R.D. (1999) The role of nutrition in crop resistance and tolerance to Imada, K., Sakai, S., Kajihara, H., Tanaka, S., and Ito, S. 2016. Magnesium oxide nanoparticles induce systemic resistance in tomato against bacterial wilt disease. Plant Pathol. 65:551-560.

Kiraly Z. 1976. Plant disease resistance as influenced by biochemical effects of nutrients and fertilizers. In Fertilizer Use and Plant Health, Proceedings of Colloquium 12. Atlanta, GA: International Potash Institute, pp. 33–46.

Knight, J. D. 1997. The role of decision support systems in integrated crop protection. Agriculture, Ecosystems & Environment, 64(2), 157-163.

Kulabhusan, P. K., Tripathi, A., and Kant, K. 2022. Gold nanoparticles and plant pathogens: An overview and prospective for biosensing in forestry. Sensors 22: 1259.

Lacoste, M. and Powles, S. 2015. RIM: Anatomy of a weed management decision support system for adaptation and wider application. Weed Science63(3): 676-689

Lawrence, R. L., Wood, S. D., & Sheley, R. L. 2006. Mapping invasive plants using hyperspectral imagery and Breiman Cutler classifications (Random Forest). Remote Sensing of Environment, 100(3), 356-362.

Lee, G., Wei, Q., and Zhu, Y. 2021. Emerging wearable sensors for plant health monitoring. Adv. Funct. Mater. 31:2106475.

Loschinkohl C., Boehm M.J. 2001. Composted biosolids incorporation improves turfgrass establishment on disturbed urban soil and reduces leaf rust severity, Hortscience 36, 790–794.

Mann R.L., Kettlewell P.S., Jenkinson P. 2004. Effect of foliar-applied potassium chloride on septoria leaf blotch of winter wheat, Plant Pathol. 53, 653–659.

Marschner H. 1995. Mineral Nutrition of Higher Plants, 2nd ed., Academic Press, London, p. 889.

Martinelli, F., Scalenghe, R., Davino, S., Panno, S., Scuderi, G., Ruisi, P., Villa, P., Stroppiana, D., Boschetti, M., Goulart, L. R., Davis, C. E., and Dandekar, A. M. 2015. Advanced methods of plant disease detection. A review. Agron. Sustain. Dev. 35:1-25

Mitter, N., Worrall, E. A., Robinson, K. E., Li, P., Jain, R. G., Taochy, C., Fletcher, S. J., Carroll, B. J., Lu, G. Q. M., and Xu, Z. P. 2017. Clay nanosheets for topical delivery of RNAi for sustained protection against plant viruses. Nat. Plants 3: 16207.

Naikoo, G. A., Mustaqeem, M., Hassan, I. U., Awan, T., Arshad, F., Salim, H., and Qurashi, A. 2021. Bioinspired and green synthesis of nanoparticles from plant extracts with antiviral and antimicrobial properties: A critical review. J. Saudi Chem. Soc. 25:101304.

Oborn I., Edwards A.C., Witter E., Oenema O., Ivarsson K., Withers P.J.A., Nilsson S.I., Richert Stinzing A. 2003. Element balances as a toll for sustainable nutrient management: a critical appraisal of their merits and limitations within an agronomic and environmental context, Eur. J. Agron. 20, 211–225.

Rad, F., Mohsenifar, A., Tabatabaei, M., Safarnejad, M. R., Shahryari, F., Safarpour, H., Foroutan, A., Mardi, M., Davoudi, D., and Fotokian, M. 2012. Detection of Candidatus phytoplasma aurantofolia with a quantum dots FRET-BASED biosensor. Jornal Plant Pathology 94:525-534.

Rajani, Mishra, P., Kumari, S., Saini, P., and Meena, R. K. 2022. Role of nanotechnology in management of plant viral diseases. Mater. Today Proc. 69:1-10.

Reid L.M., Zhu X., Ma B.L. 2001. Crop rotation and nitrogen effects on maize susceptibility to gibberella (Fusarium graminearum) ear rot, Plant Soil 237, 1–14.

Reuveni M., Agapov V., Reuveni R. 1997b. Controlling powdery mildew caused by Sphaerothecafuliginea in cucumber by foliar sprays of phosphate and potassium salts. Crop Prot. 15: 49–53.

Reuveni M., Oppernheim D., Reuveni R. 1998. Integrated control of powdery mildew on apple trees by foliar sprays of mono-potassium phosphate fertilizer and sterol inhibiting fungicides, Crop Prot. 17, 563–568.

Reuveni M., Oppernheim D., Reuveni R. 1998. Integrated control of powdery mildew on apple trees by foliar sprays of mono-potassium phosphate fertilizer and sterol inhibiting fungicides. Crop Prot. 17, 563–568.

Reuveni R., Dor G., Raviv M., Reuveni M., Tuzun S. 2000. Systemic resistance against Sphaerothecafuliginea in cucumber plants exposed to phosphate in hydroponics system, and its control by foliar spray of mono-potassium phosphate. Crop Prot. 19,355–361.

Robinson P.W., Hodges C.F. 1981. Nitrogen-induced changes in the sugars and amino acids of sequentially senescing leaves of Poa pratensis and pathogenesis by Drechslerasorokiniana, Phytopathol. Z. 101, 348–361.

Römheld V., Marschner H. 1991. Function of micronutrients in plants, in: Mortvedt J.J., Cox F.R., Shuman L.M., Welch R.M. (Eds.), Micronutrients in Agriculture. Soil Science Society of America, Inc. Madison, Wisconsin, USA, pp. 297–328.

Rosa, C., Kuo, Y.-W., Wuriyanghan, H., and Falk, B. W. 2018. RNA interference mechanisms and applications in plant pathology. Annu. Rev. Phytopathol. 56: 581-610.

Sabry, A.-K. H. 2019. Role of nanotechnology applications in plant-parasitic nematode control. Pages 223-240 in: Nanobiotechnology Applications in Plant Protection. Nanotechnology in the Life Sciences. K. Abd-Elsalam and R. Prasad, eds. Springer, Cham, Switzerland.

Santana, I., Jeon, S.-J., Kim, H.-I., Islam, M. R., Castillo, C., Garcia, G. F. H., Newkirk, G. M., and Giraldo, J. P. 2022. Targeted carbon nanostructures for chemical and gene delivery to plant chloroplasts. ACS Nano 16: 12156-12173

Schmidt, J.P. 2011. Why Are Some Plant Genera More Invasive Than Others? Plos One 6(4): e18654

Shan, P., Lu, Y., Lu, W., Yin, X., Liu, H., Li, D., Lian, X., Wang, W., Li, Z., and Li, Z. 2022. Biodegradable and light-responsive polymeric nanoparticles for environmentally safe herbicide delivery. ACS Appl. Mater. Interfaces 14: 43759-43770.

Shang, H., Ma, C., Li, C., White, J. C., Polubesova, T., Chefetz, B., and Xing, B. 2020. Copper sulfide nanoparticles suppress Gibberellafujikuroi infection in rice (Oryza sativa L.) by multiple mechanisms: Contactmortality, nutritional modulation and phytohormone regulation. Environ. Sci. Nano 7:2632-2643.

Sharma R.C., Duveiller E. 2004. Effect of helminthosporium leaf blight on performance of timely and late-seeded wheat under optimal and stressed levels of soil fertility and moisture, Field Crop Res. 89, 205–218.

Sharma S., Duveiller E., Basnet R., Karki C.B., Sharma R.C. 2005. Effect of potash fertilization on helminthosporium leaf blight severity in wheat, and associated increases in grain yield and kernel weight, Field Crop Res. 93, 142–150.

Shidore, T., Zuverza-Mena, N., White, J. C., and da Silva, W. 2021. Nanoenabled delivery of RNA molecules for prolonged antiviral protection in crop plants: A review. ACS Appl. Nano Mater. 4:12891-12904.

Simoglou K., Dordas C. 2006. Effect of foliar applied boron, manganese and zinc on tan spot in winter durum wheat, Crop Prot. 25, 657–663.

Small, I. M., Joseph, L., & Fry, W. E. 2015. Evaluation of the BlightPro decision support system for management of potato late blight using computer simulation and field validation. Phytopathology, 105(12), 1545-1554.

Sønderskov, M., Rydahl, P., Bøjer, O. M., Jensen, J. E., & Kudsk, P. 2016. Crop protection online-weeds: a case study for agricultural decision support systems. Real-World Decision Support Systems: Case Studies, 303-320.

Sousa, G. F. M., Gomes, D. G., Campos, E. V. R., Oliveira, J. L., Fraceto, L. F., Stolf- Moreira, R., and Oliveira, H. C. 2018. Post-emergence herbicidal activity of nanoatrazine against susceptible weeds.

Stone A.G., Scheuerell S.J., Darby H.D. 2004 Suppression of soilborne diseases in field Thangavelu, R. M., Kadirvel, N., Balasubramaniam, P., and Viswanathan, R. 2022.

Timonin M.E. 1965. Interaction of higher plants and soil microorganisms. in: Gilmore C.M., Allen O.N. (Eds.), Microbiology and Soil Fertility. Oregon State University Press, Corvallis, OR, pp. 135–138.

Tziros G.T., Dordas C., Tzavella-Klonari K., Lagopodi A.L. 2006. Effect of two Pseudomonas strains and Fusarium wilt of watermelon under different nitrogen nutrition levels, Proceedings of the 12th Congress of Mediterranean Phytopathological Union, 11–15 June 2006, Rhodes Island Hellas, pp. 585–587.

Ultrasensitive nano-gold labelled, duplex lateral flow immunochromatographic assay for Wagi, S., and Ahmed, A. 2020. Bacterial nanobiotic potential. Green. Process. Synth. 9: 203-211.

Wang, Q., Zhang, C., Shen, G., Liu, H., Fu, H., and Cui, D. 2014. Fluorescent carbon dots as an efficient siRNA nanocarrier for its interference therapy in gastric cancer cells. J. Nanobiotechnol. 12:58.

Wiese J., Bagy M.M.K. and Shubert S. 2003. Soil properties, but not plant nutrients (N, P, K) interact with chemically induced resistance against powdery mildew in barley. J. Plant Nutr. Soil Sci. 166: 379– 384

Worrall, E. A., Bravo-Cazar, A., Nilon, A. T., Fletcher, S. J., Robinson, K. E., Carr, J. P., and Mitter, N. 2019. Exogenous application of RNAi-inducing double-stranded RNA inhibits aphid-mediated transmission of a plant virus. Front. Plant Sci. 10:265.

Xing, B. 2021. Elemental sulfur nanoparticles enhance disease resistance in tomatoes.

Yao, P., Zou, A., Tian, Z., Meng, W., Fang, X., Wu, T., and Cheng, J. 2021. Construction and characterization of a temperature-responsive nanocarrier for imidacloprid based on mesoporous silica nanoparticles. Colloids Surfaces B Biointerfaces 198:111464.

3

Geospatial Vigilance: A Crop Health Assessment Tool to Forecast Disease Outbreaks in Sustainable Agriculture

Sibte Sayyeda[1], Vaibhav Pratap Singh[1], Devesh Pathak[2] and Deepak Mourya[3]

[1]Department of Plant Protection, Faculty of Agricultural Sciences Aligarh Muslim, University, Aligarh-202002, Uttar Pradesh
[2]Department of Plant Pathology, RSM College, Dhampur, Bijnor-246761 Uttar Pradesh
[3]Department of Plant Pathology, Shri Khushal Das University Hanumangarh-335801, Rajasthan

Abstract

Food security is paramount for developing any nation reliant upon sufficient crop production. The decline in crop production significantly impacts food security, leading to an imbalance between global food demand and agricultural output. Multiple factors contribute to reduced agricultural productivity, including abiotic factors like water scarcity, poor soils, and unsuitable temperatures, as well as biotic factors such as pests, diseases, and weeds attacking crops. These challenges result in inefficient input utilization and decreased crop yields, ultimately threatening food security. Pests, diseases, and weeds profoundly impact plant health, causing substantial losses in crop production. Geospatial technology is employed throughout the agricultural cycle, from initial crop health assessment to forecasting disease outbreaks. Despite the significant challenges global and local agriculture face in achieving optimal outputs and securing food for the world population, the wealth of geospatial data and technological advancements plays a crucial role. They aid decision-makers in formulating strategies to combat various pests and diseases affecting plant health and food crops, thereby contributing to improved agricultural outcomes and food security. This chapter provides a comprehensive appraisal of existing literature to elucidate the role of geospatial technologies, encompassing Geographic Information System (GIS), remote sensing, and the Global Navigation Satellite System (GNSS), which are utilized for data collection, mapping, analyzing distribution, and predicting events related to plant health.

Keywords: *Geospatial technology, GPS, Geographic Information System (GIS), Remote Sensing, Global Navigation Satellite System (GNSS), pests and diseases, management*

Introduction

Early childhood malnutrition is a significant concern due to its potential long-term effect on child development and a country's economic growth (Regmi and Paudel, 2017). In developing nations, about one-third of children experience underweight or stunted growth, with undernutrition causing over one-third of deaths among children under five years old. This undernutrition often includes micronutrient deficiencies, known as hidden hunger, resulting from inadequate dietary intake and diseases. These issues stem from food insecurity, insufficient access to safe and clean water, poor maternal and childcare practices, and inadequate health services and sanitation (Tirado *et al.*, 2013). Additionally, there has been a growing disparity between the increasing global food demand and agriculture output since 2008. This mismatch between supply and demand at local, regional, and national levels has exacerbated the situation further. These factors collectively contribute to the challenge of addressing malnutrition and its consequences, particularly in developing countries (Savary *et al.*, 2012).

Numerous factors can significantly impact agricultural productivity. Both abiotic and biotic constraints, such as water scarcity, poor soil quality, unfavorable temperatures, and various crop pests, diseases, and weeds, contribute to reduced crop yields; this leads to inefficient use of inputs, lower crop output, and, ultimately, food insecurity (Knox *et al.*, 2012). Climate change has a profound negative effect on global agricultural productivity, especially in tropical zones (Masters *et al.*, 2010). The ongoing climate disturbances are anticipated to lead to a 20% reduction in agricultural production by 2080, a particularly concerning trend for developing countries like India. These climate shifts result in extreme events such as heat and cold stresses, erratic rainfall patterns, droughts, floods, storms, etc., significantly impacting crop yields. Furthermore, these environmental changes contribute to the emergence of new insect pests and disease outbreaks, further challenging agricultural sustainability (Masters *et al.*, 2010).

Since the establishment of the World Trade Organization (WTO) in 1995, there has been a notable increase in global trade of agricultural products due to trade liberalization. This globalization of trade has facilitated the movement of invasive alien species, disrupting local and exotic plant ecosystems in various regions. Both trade liberalization and climate change have also contributed to the emergence of new diseases and insect pest outbreaks (Masters and Norgove, 2010). However, advancements in technology offer ways to minimize crop yield losses. Geographical Information System (GIS) and Remote Sensing are now widely used to accurately collect crop information and analyze data, playing a crucial role in addressing the effects of diseases and insect pests on plants/crops. This chapter aims to explore the valuable role of geospatial technology in tackling these challenges in agriculture and provide insights into its implementation to mitigate these issues.

Challenges Posed by Pests and Diseases on Plant Health

When a plant operates at its optimal genetic potential, it is considered as healthy. However, when plants face insect infestations or pathogen attacks, their normal physiological functions are disturbed (Azoulay, 2017), reducing plant health, biomass, productivity, and overall production. The damage caused by insect pests and diseases can occur in several ways:

- **Direct kill-off** – Some pests or diseases can directly kill and affect neighboring plants. Examples include soil fungi, vascular diseases, breakdowns, and certain boring insects.
- **Stunted growth** – Other disruptions can lead to stunted growth due to metabolic issues, nutritional deficiencies, or root damage. For instance, aphids and various plant viruses can cause such problems.
- **Branch damage** – Certain pests and diseases target and kill specific parts of plants, such as branches. This can occur with fungal diebacks and certain boring insects.
- **Leaf damage** – Lastly, some pests and diseases specifically target and damage leaf tissues, causing issues like leaf spots, blights, mildews, rusts, and damage from certain boring insects.

Insect pests or diseases typically attack plants in the field but become more noticeable after harvest, affecting yield and quality. The cumulative impact of these pests and diseases can be severe, harming agricultural produce and significantly affecting forest health and ecosystem services (Potter and Urquhart, 2017). Forest diseases and pests can alter natural landscapes, cultural values, wildlife habitats, biodiversity, and the forest's ability to sequester carbon, ultimately impacting the economy of forest-based communities and the sustainability of forests (Trevor *et al.*, 2012; Boyd *et al.*, 2013; Anwer and Singh, 2019). Therefore, addressing these challenges is crucial for maintaining crop productivity and forest ecosystems' overall health and economic viability.

Impact of Pests and Diseases on Agricultural Crop Health

Pests in agriculture are organisms responsible for causing harm or destruction to plants (Jolivet, 1988). This category primarily includes insect pests. These pests can cause damage in various ways, such as through sucking, boring, cutting, and other mechanisms, and they belong to different species across various orders. Agricultural crops worldwide face threats from over 100,000 diseases caused by fungi, bacteria, and other microorganisms, as well as from approximately 30,000 species of weeds, 10,000 species of insects, and 1,000 species of nematodes (Dhaliwal *et al.*, 2010). However, only a tiny fraction, less than 10 percent, of these identified pest species are typically considered major pests. The dynamics of insect pest problems in agriculture have undergone significant changes in the twenty-first century's first decade due to shifts in ecosystems and technological advancements.

In contrast, there has been a decrease in the severity of certain pests like *Helicoverpa armigera,* which can severely impact crop productivity and production. This pest can multiply up to eleven generations annually under ideal conditions (Pandey *et al.*, 2016). Other pests have shown an increasing trend in incidence; examples include mealy bugs, particularly *Phenacoccus solenopsis*, affecting cotton crops; the sugarcane woolly aphid, *Ceratovacuna lanigera*, on sugarcane; the tobacco caterpillar, *Spodoptera litura*, affecting various crops; and the diamondback moth, *Plutella xylostella*, remaining highly destructive to cruciferous vegetables. One of the most devastating pests in recent times is the fall armyworm (*S. frugiperda*), especially for maize crops. This pest has caused significant economic yield losses ranging from 22% to 67% in Ghana and Zambia (Day *et al.*, 2017) and 32% to 47% in Ethiopia and Kenya (Kumela *et al.*, 2018). In Africa and Asia, *S. frugiperda* generations have been observed continuously throughout the year, affecting host plants, including off-season and irrigated crops (Prasanna *et al.*, 2018). The harm caused by pests to agricultural crops is multifaceted. Pests can host on plants, nest in fields, or embed eggs in plant tissues, leading to damage. This damage can reduce leaf surface area, hindering water transport and nutrient uptake. Stem damage can further impair water and nutrient flow, while root damage affects mineral and water absorption from the soil, ultimately slowing plant growth and promoting tissue rot.

Identifying the impact of pests on plants often involves observing pathological changes, such as spots on plant surfaces, which serve as indicators of damage. Sucking pests, for instance, damage plants by extracting juice using their mouthparts, contributing to plant health decline and growth inhibition (James and Clay, 2016). Furthermore, the presence of these pests alters the biochemical composition of plant cells and tissues due to the secretion injected by their salivary glands while feeding. Consequently, heavily infested plants exhibit symptoms such as wilting, yellowing, stunted growth, deformities, and eventual death. Some sucking insects also introduce toxic substances and carry disease-causing organisms, further compounding the damage they cause. Common examples of sucking pests include aphids, blackflies, greenflies, whiteflies, ticks, true bugs, and thrips.

India's diverse agricultural landscape faces specific pest challenges for each crop. For instance, brinjal (eggplant) is vulnerable to attacks by over 70 insect species (Subbarathnam and Butani, 1982). Notable pests include the brinjal shoot and fruit borer (BSFB) (*Leucinodes orbonalis*), aphid (*Aphis gossypii*), leafhopper (*Amraska bigutulla*), stem borer (*Euzophera perticella*), whitefly (*Bemisia tabaci*), Epilachna beetle (*Henosepilachna vigintioctopunctata*), and lacewing bug (*Urantitus hystricellus*), with BSFB alone, causing 80-90% yield loss (Patnaik, 2000; Misra, 2008; Jagginavar *et al*., 2009). Additionally, the diamondback moth (*Plutella xylostella*) is a significant pest affecting cabbage and cauliflower (Robin *et al*., 2017), belonging to the Plutellidea family and *Plutella genus.* The *S. frugiperda* has become a serious threat to Indian maize crops since 2018. Moreover, oil palms face potential damage from around 80 species of arthropods that are pests capable of feeding on them. Each country faces unique challenges with pests and diseases that can significantly impact agricultural activities and lead to economic loss. For instance, in America, pests like *Rhynchophorus palmarum* and *Alurnus humeralis* attack palm oil crops, while in India, the population of coccimellidae in the spear region poses a threat. In Kerala, *Rhynchophorus ferrugineus* is a recorded pest (Kalidas, 2012). These pests contribute to substantial economic losses in agriculture. The global losses due to insect pests have been recorded at 10.8 percent from the post-green revolution era to the beginning of this century. In India, crop losses have reached 17.5 percent during the same period, resulting in an annual loss of about Rs. 863.884 billion due to insect pests (Dhaliwal *et al*., 2010).

Plant diseases, which can be caused by living (biotic) or non-living (abiotic factors), also contribute significantly to agricultural losses. Biotic factors include fungi, bacteria, phytoplasmas, viruses, viroid, virus-like organisms, nematodes, parasitic plants, and protozoa (Busby *et al*., 2016). In contrast, abiotic factors include heat storms, frost, wind, soil salt, soil compaction, and root damage or girdling (Alfren, 2001). Among these, fungal pathogens are destructive and can cause significant crop losses (Alfren, 2001). An example of a devastating plant disease is the late blight of potato (*Phytophthora infestans*), which resulted in the Irish famine of 1845/46, causing significant population loss.

Fungi lack conductive tissue and chlorophyll and are classified as parasites or saprophytes because they rely on external sources for nutrition, often decaying organic matter. They can harm plants by killing cells or inducing stress. For example, *Synchytrium endobioticum* causes potato wart disease and was considered a potential bioterrorism agent in the U.S. in 2002 (Nutter and Madden, 2009). Another example is that the Moko disease of banana caused by Ralstonia solanacearum (Blomme *et al*., 2017) is a significant environmental and economic threat, causing substantial losses in many tropical regions. Neglecting disease control measures can lead to severe economic consequences. Based on economic losses, the 'top ten' fungal pathogens are ranked as follows: *Magnaporthe oryzae, Botrytis cinerea, Puccinia* spp., *Fusarium graminearum, F. oxysporum, Blumeria graminis, Mycosphaerella graminicola, Colletotrichum* spp., *Ustilago maydis, and Melampsora lini.* Honorable mentions go to *Phakopsora pachyrizi* and *Rhizoctonia solani,* which just missed the top ten (Dean *et al*., 2012). These phytopathogenic fungi are commonly found in warm and humid conditions.

Abiotic stresses in plants occur due to abiotic factors such as wind, frost, soil compaction, etc., and they cannot spread from affected plants to healthy plants (Zandalinas *et al*., 2018). These diseases often result from unfavorable environmental conditions that disrupt normal plant growth. Factors like inadequate or excessive levels of essential physical or chemical elements necessary for healthy plant development can lead to abnormal growth patterns. For instance, low humidity levels can increase evaporative demand, causing stress on plant moisture even with sufficient root water supply, resulting in tissue damage. Also, atmospheric gases, temperature variations, and light intensity can contribute to abiotic stresses (Crandall and Gillbert, 2017). Understanding and managing these environmental factors are essential for preventing and mitigating abiotic stresses in plants.

Geospatial Innovations Revolutionizing Agricultural Practices

Geospatial technology comprises computer hardware, software, geographical data, and skilled personnel aimed at capturing, storing, updating, manipulating, analyzing, and presenting geographically referenced information efficiently. It encompasses geographical information system (GIS), global positioning systems (GPS), and remote sensing (RS), aiding users in collecting, analyzing, and interpreting spatial data. This technology explores the relationship and statuses of human-made and natural objects within Earth's space and beyond. Geospatial technology has gained significant traction in India's public and private sectors, spanning various industries such as agriculture, telecommunications, oil and gas, forestry, environmental management, public safety, infrastructure, and logistics. As stakeholders recognize the value and cost-effectiveness of geospatial tools and technologies, the industry is poised for exponential growth in the coming years.

The application of geospatial technology in agriculture dates back to the early 1990s (Sood *et al*., 2015), marking a significant advancement. Farmers can now leverage global navigation satellite system (GNSS) technology to measure spatial and temporal variability in soil, relief, and vegetation. This includes locating agricultural machinery efficiently and enhancing the quality and accessibility of geographic information in digital formats. Utilizing mobile technology, farmers can conveniently manage and retrieve farm and field records on- site, streamlining data entry and access (Qu and Tao, 2014). GPS and GIS technology are vital in creating geo-referenced maps that provide insights into different crops, soil properties, diseases, and insect pests. These tools offer field professionals and growers enhanced communication and management capabilities (Deleon *et al*., 2017). The diverse range of map production available today improves insect pests, diseases, and crop management practices. For instance, a soil map offers valuable insights into land properties, aiding in better land management decisions. The inherent spatial and temporal references associated with agricultural parcels, including yield data and operational records like fertilization and ploughing, further highlight the utility of geospatial technology in enhancing agricultural practices and productivity.

Mapping India's Geospatial Advancements

India's GDP is expected to reach $9-10 trillion by 2050, shifting towards a highly industrialized and technologically advanced economy. India needs a robust information and knowledge infrastructure to prepare for this growth. Geospatial technologies will play a pivotal role in managing information efficiently. Their applications carry significant social and national implications, supporting governance, sustainable development planning, business process optimization, and citizen access to geographical knowledge. The geospatial sector in India is projected to exceed Rs 63,000 crore by 2025, growing at a rate of 12.8% annually. This growth is anticipated to create employment opportunities for over 1 million individuals, with a significant focus on fostering Geospatial Start-Ups (UN-WGIC, 2022). States like Andhra Pradesh, Karnataka, Rajasthan, and Tamil Nadu already leverage geospatial technology for effective governance and streamlined management practices.

Harnessing Geographic Information System for Crop Health, Pest and Disease

Management

Geographic Information System (GIS) are instrumental in storing, capturing, analyzing, and visualizing data related to specific parts of the Earth's surface, encompassing technical, administrative, and scientific aspects like geoscience, economics, and ecological application (Murray, 2017). GIS functionalities typically include data retrieval and storage, visualization, querying, geometric and thematical analysis, and transformation (Barleme, 2011). It can be segmented into four categories: data capture, analysis, administration, and presentation systems. With the rapid advancement of technology, numerous options are available for data capture, such as global positioning system (GPS), geodetic surveying, laser scanning, photogrammetry, and satellite-based remote sensing. Consequently, the evolution of geospatial technology has significantly simplified plant health monitoring.

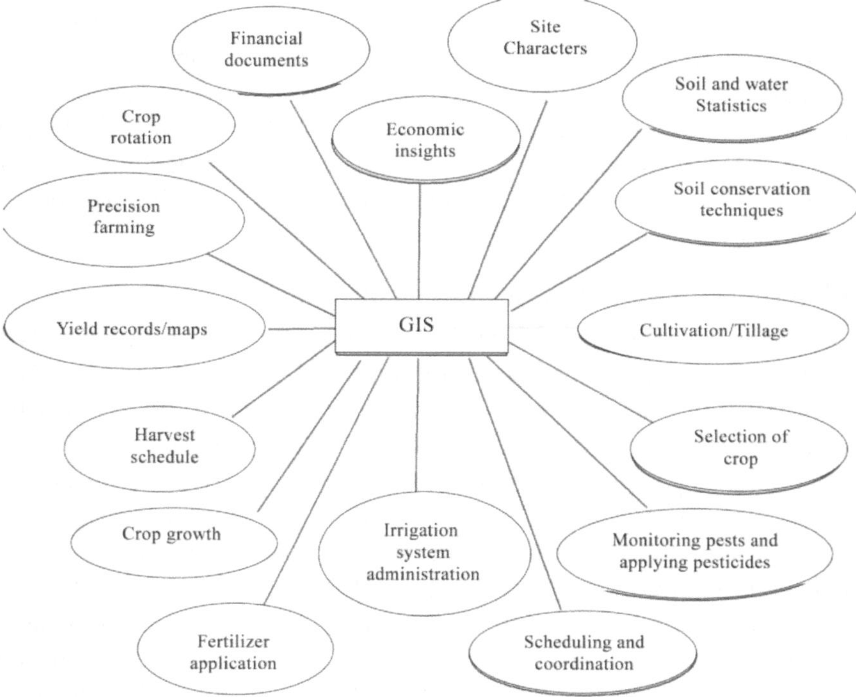

Fig. 1: Implementing GIS across multiple farming tasks

GIS is extensively utilized in numerous agricultural activities, particularly in facilitating biosecurity evaluations related to plant health (Figure 1)(Lindgren,

2012). Mandal *et al.* (2013) highlighted an instance where Brazil tackled the challenge of up to 40% loss in monoculture cotton production due to *Rotylenchulus reniformis* infestation by employing geostatistical techniques to generate a spatially differentiated nematode risk map within the field. This approach is more cost-effective than relying solely on nematicide for protection.

GIS is also applied to enhance input management practices among farmers in New Zealand (Meeta, 2017). Ravensdown, a prominent manufacturer and distributor of fertilizer in the country, employs geospatial technology to assist farmers in reducing the wastage of resources that could lead to harmful runoff into streams and water bodies. This approach benefits the environment and results in potential savings of up to 10% per year in total fertilizer expenditure for farmers.

Furthermore, GIS is recognized as a software application for managing and analyzing spatial data, encompassing aspects related to productivity, agronomy, disease, and pests. Case studies conducted in India have demonstrated that implementing a GIS-based Decision Support System can assist farmers in making informed day-to-day decisions. Agricultural experts leverage this system to recommend crop selections based on geographic considerations, considering various risks. Additionally, GIS aids in analyzing historical records or databases in conjunction with geographical maps, facilitating the creation of diverse models for agricultural practices. In essence, GIS serves as a valuable tool for decision-makers by identifying existing cultivation sites and analyzing the potential of other sites for various agricultural purposes, such as floriculture (Manas, 2017).

For semi-automated data analysis, GIS/GPS technology was employed in northern California, Washington, Oregon, Arizona, and Idaho. This approach integrated ground-based weather observations, plant-stage measurements, and remote imagery, all geo-referenced within GIS software. The aim was to assess insect and disease risk and determine crop cultural requirements. Elevation, weather, and satellite data were combined to develop weather and disease weather forecasts specifically tailored to agricultural regions. Geo-referenced maps, updated daily, displayed models for six insect pests and twelve crop diseases across surveyed agricultural areas. Additionally, highly accurate predictions were made for grape harvest dates and yields. The data generated from GIS/GPS analyses guided management decisions across vast acreage in the study area, and this information was disseminated daily over the internet as regional maps showcasing weather conditions, insect risks, and disease probabilities (Thomas *et al.*, 2002).

In Malaysia, Pohl *et al.* (2016) utilized GIS to support decision-making and monitoring, aiming to enhance the sustainability of oil palm plantations. After geo-coding, they integrated spatial layers from various sources into GIS, followed by employing suitable GIS methods to analyze the collected data. Similarly, Yahaya *et al.* (2015) also applied GIS in managing paddy cultivation. Their research led to the development a web-based system for monitoring paddy growth. All gathered data were converted into spatial information and stored, culminating in creating a map published on the ArcGIS server. This database is accessible via internet information systems through web browsers, with ArcGIS servers used to publish spatial layers as map series.

Integrating Remote Sensing in Disease and Pest Management Strategies

Remote sensing involves acquiring information about an object without direct contact. It relies on electromagnetic radiation as the information carrier, traveling through a vacuum as waves at the speed of light. There are two main types: active and passive remote sensing. Passive remote sensing sensors detect incident rays reflected or emitted from objects. In contrast, active sensors emit radiation that interacts with the target and return to the measuring device for analysis (Lillesand, 2014). In agriculture, spectrometers are commonly used in remote sensing to address various challenges like plant nutrition, disease, pest detection, yield prediction, water management, and weed control. Figure 2 illustrates the factors conducive to pest development and their impact on detectable plant parameters, as observed by remote sensors.

Remote sensing technology finds applications across various domains such as environmental monitoring, site-specific agronomic management (SSM), climate and land-use change detection, land cover classification, drought, disease, and pest monitoring. Its capability to discern minute differences in vegetation renders it valuable for quantifying within-field variability, assessing crop growth, and implementing field management strategies based on real-time conditions, aspects that might be missed through conventional ground-based visual scouting methods.

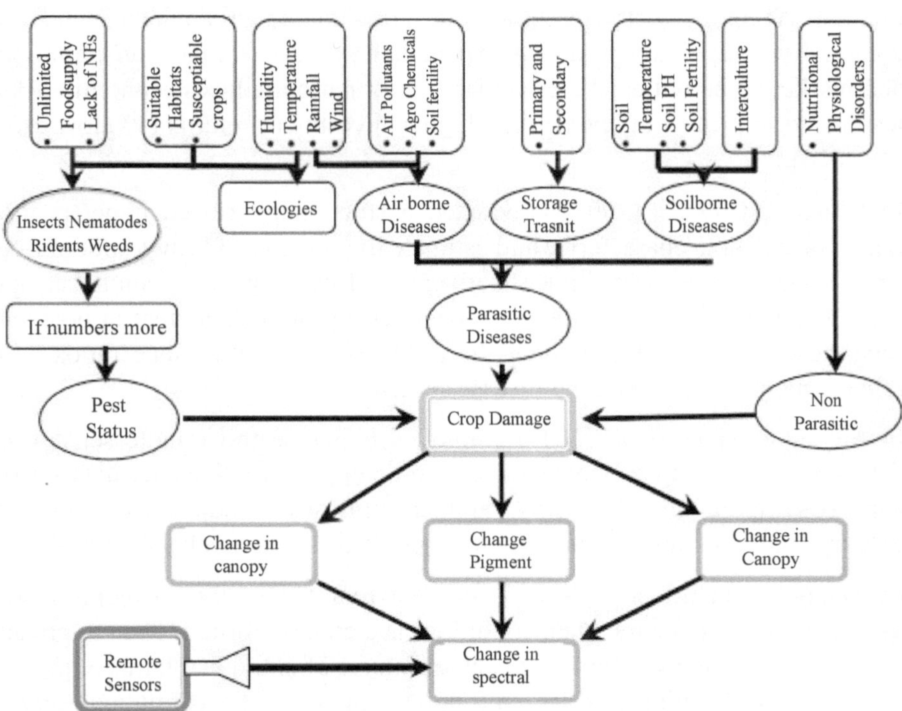

Fig. 2: Remote sensing techniques for identifying, monitoring, and mitigating crop damage caused by pests and diseases (Source: Anwer and Singh, 2019).

Wojtowicz *et al.* (2016) highlighted the use of remote sensing in determining the nutritional needs of plants. They conducted a study involving two sets of cucumber plants (*Cucumis sativus*) grown under controlled conditions. One set was inoculated with the bacterial wilt pathogen (*Ralstonia solanacearum*), while the other served as the control. During the pre- symptomatic stage, both sets underwent changes relevant to field conditions, such as alterations in nutrient levels, water content, and light exposure. They also measured the spatial reflectance of the cucumbers using a spectro-radiometer. In this study, remote sensing instruments detect electromagnetic radiation reflected or emitted by objects of interest. If two objects reflect the same electromagnetic energy, they appear identical in remotely sensed images. However, when plants are infected by pathogens like fungi, bacteria, and viruses, there may be an initial period without visible symptoms. This could be due to the plant's resistance mechanism or the time taken by the pathogen to establish itself within the plant (Elibatty *et al.*, 2014).

Moreover, remote sensing has been employed to predict crop yields, primarily relying on statistical-empirical correlations between yield and vegetation indices. Accurate yield forecasts are crucial for government agencies, commodity traders, and producers to plan their harvest, storage, transportation, and marketing activities. Access to such information reduces economic risks, leading to improved efficiency and higher investment returns (Wojtowicz *et al.*, 2016).

In China, Liu *et al.* (2008) conducted a study that utilized hyperspectral reflectance data to characterize and estimate the severity of brown spot of rice. The findings demonstrated the effectiveness of hyperspectral remote sensing data in characterizing crop diseases for precise pest management in practical scenarios. They also highlighted the utility of crop reflectance ratios as a reliable method for estimating disease severity.

In Australia, Apan *et al.* (2005) employed hyperspectral remote sensing to detect pests and diseases in vegetable crops, comparing reflectance data across various symptom sets. Similarly, another study was conducted in the same country using the same technology to detect leaf rust in blueberries early.

In Malaysia, Tayari *et al.* (2015) utilized remote sensing to monitor soil usage, capturing dynamic plant conditions in a concise format. Their approach involved measuring visible and invisible features of individual farms or groups of farms, transforming this data into point estimations for continuous spatial information.

The Food and Agricultural Organization (FAO) has created an android application, FAMEWS, designed to record field scouting and pheromone trap data. Ongoing research in remote sensing, automatic counting of trap catches, image analysis of insects and their damage, and radar technology will significantly enhance monitoring efforts and contribute to a deeper understanding of fall armyworm (*Spodoptera frugiperda*) biology. These advancements also present opportunities for developing forecasting methods related to fall armyworm infestations.

Remote sensing technology is applicable in diverse studies, including precision farming in rice field management. For instance, Pandak *et al.* (2014) utilized high-resolution satellite imagery ranging from 0.5 m to 1.0 m, overlaid with an integrated cadastral map. This approach helped accurately delineate the planted areas and distinguish between paddy fields and the non-paddy regions, such as roads and fish ponds. The study primarily aimed to leverage high-resolution images for effective rice field management. Satellite imagery has been instrumental in mapping tasks and has facilitated monitoring and observation of activities throughout the crop growth cycle, from land preparation to harvest.

Leveraging Global Navigation Satellite System (GNSS) for Pest and Disease

Management

The Global Navigation Satellite System (GNSS) has been extensively utilized for various applications, including location services, navigation, tracking, mapping, and timing. The Global Positioning System (GPS) is the most widely known GNSS, developed in the 1980s and fully operational by 1995. Ground-receiving units can simultaneously capture location data from multiple satellites, enabling triangulation to determine precise receiver locations (Thompson, 1998). GNSS technology has gained importance in agriculture, supporting research and practical applications. Accurate, up-to-date, spatial, or non-spatial information is crucial in agricultural management for efficient planning and execution, aiming to enhance land productivity and input utilization (Palaniswami *et al*., 2011). Precision farming extensively employs GNSS technology to gather geo-referenced data points and more. Typically, Differential GNSS (DGNSS) or Real-Time Kinematic GNSS (RTK-GNSS) is preferred due to their submeter-level accuracy. In managing plant pests and diseases, Jiannong *et al.* (2009) utilized advanced digital imagining technology integrated with GPS coordinates and timestamps to create geo-tagged images. This technique facilitates the monitoring and mapping pest and disease occurrences in plants over time and space. Geo- coded images can be seamlessly transferred to a computer using metadata like Exchangeable Image File Format (EXIF) for integration into a mapping system. The mapping process is streamlined by integrating mapping software and merging web mapping with digital imaging technologies. This integration enhances tracking pests and diseases across different temporal and spatial dimensions.

In the realm of plant pest and disease management, Gaugherty *et al.* (2015) utilized GPS technology to pinpoint areas positive for plum pox virus (PPV) alongside PPV-negative prunus block. PPV is a highly destructive pathogen, significantly impacting fruit yield and quality. Similarly, in combating the Panama wilt of banana in the Middle East, Ploetz *et al.* (2015) employed GPS to identify infected and non-infected areas, mapping the data for analysis. Aguilar *et al.* (2014) applied a geo-location method with GPS technology for in situ habitat characterization in addressing banana insect pests in Mindanao, Philippines. Over six months, they monitored insect pest incidence and compared locations to analyze crop infestations. Likewise, in Malaysia, Idris *et al.* (2015) used handheld GPS devices for data collection, identifying locations affected by pests and diseases.

Conclusion

Geospatial data and technology are pivotal in enhancing crop production and addressing food security concerns. As the global population expands, there is a pressing need to increase crop production to ensure ample nutrition for everyone. This chapter explains the role of geospatial technologies in addressing pest and disease challenges that significantly impact crop yield, which is crucial for sustaining the world's food needs. Geospatial technology has been instrumental, starting from assessing crop health status to managing crop data. Many studies have utilized remote sensing technology to monitor plant health for major crops like oil palms, paddy, and wheat, critical commodities for countries. However, due to its high costs, this technology is limited to small-scale crops such as orchards.

Additionally, detecting plant health for randomly planted crops, especially in mixed land use, is challenging via satellite imagery. GPS technology complements ground crews during field surveys by automatically recording the locations of affected plants. Mapping the distribution of pest and disease incidence helps visualize these events on a large scale. Spatial analysis aids in predicting future incidence possibilities, contributing to proactive management strategies. Global trade liberalization and climate change significantly challenge local crop production and food security. Nevertheless, advancements in geospatial technology have streamlined the process of combating various pests and diseases affecting plant health compared to previous methods.

References

Aguilar, C. H., Lasalita-Zapico, F., Namocatcat, J., Fortich, A., & Bojadores, R. M. (2014). Farmers' perceptions about banana insect pests and integrated pest management (IPM) systems in SocSarGen, Mindanao, Philippines. IPCBEE, 63(5), 22-27.

Alfen, N. K. V. (2001). Fungal pathogens of plants, ed K. Roberts (Encyclopedia of life science: Wiley inter science, Chichester).

Anwer, A., & Singh, G. (2019). Geo-spatial technology for plant disease and insect pest management. Bulletin of Environment, Pharmacology and Life Sciences, 8(12), 1-12.

Apan, A., Datt, B., & Kelly, R. (2005, January). Detection of pests and diseases in vegetable crops using hyperspectral sensing: a comparison of reflectance data for different sets of symptoms. In Proceedings of the 2005 Spatial Sciences Institute Biennial Conference 2005: Spatial Intelligence, Innovation and Praxis (SSC2005).

Barleme N. (2011). Geographic information system Springer Handbook of Geographic Information ed Kresse W. & Danko D. (Springer: Berlin, Heidelberg).

Blomme, G., Dita, M., Jacobsen, K. S., Pérez Vicente, L., Molina, A., Ocimati, W., & Prior, P. (2017). Bacterial diseases of bananas and enset: current state of knowledge and integrated approaches toward sustainable management. Frontiers in plant science, 8, 1290.

Busby, P. E., Ridout, M., & Newcombe, G. (2016). Fungal endophytes: modifiers of plant disease. Plant molecular biology, 90, 645-655.

Crandall, S. G., & Gilbert, G. S. (2017). Meteorological factors associated with abundance of airborne fungal spores over natural vegetation. Atmospheric Environment, 162, 87-99.

Dean, R., Kan, J. A. V., Zacharias, A., Pretorius, Z. A., Hammond-Kosack, K. E., Di Pietro, A., Spanu, P. D., Rudd, J. J., Dickman, M., Kahmann, R., Ellis, J., & Foster G. D. (2012). The Top 10 fungal pathogens in molecular plant pathology. Molecular plant pathology, 13(4), 414-430.

Deleon, L., Brewer, M. J., Esquivel, I. L., & Halcomb, J. (2017). Use of a geographic information system to produce pest monitoring maps for south Texas cotton and sorghum land managers. Crop protection, 101, 50-57.

Elbattay, A., Shiuan, Y. T., Sijam, K., & Hashim, M. (2014). Induced Em-Spectral Response for an Early Detection of Vegetation (Crop) Biotic Stress Using Hyperspectral Spectroradiometer. Arab Remote Sensing and Geographic Information System Corporation, 81-84.

Gougherty, A. V., & Nutter Jr, F. W. (2015). Impact of Eradication Programs on the Temporal and Spatial Dynamics of Plum pox virus on Prunus spp. in Pennsylvania and Ontario, Canada. Plant Disease, 99(5), 593-603.

Idris, N. H., Said, M. N., Fauzi, M. F., Yusri, N. A. M., & Ishak, M. H. I. (2015). A Low Cost Mobile Geospatial Solution to Manage Field Survey Data Collection of Plant Pests and Diseases. In IEEE Workshop on Geoscience and Remote Sensing, Kuala Lumpur Malaysia: IWGRS.

Jagginavar, S. B., Sunitha, N. D., & Biradar, A. P. (2009). Bioefficacy of flubendiamide 480 SC against brinjal fruit and shoot borer, Leucinodes orbonalis Guen. Karnataka Journal of Agricultural Sciences, 22(3), 712-713.

James, F.D., & Clay A. K. (2016). Umaine Cooperative Extension: Insect Pests, Ticks and Plant Disease [online] Available: https://extension.umaine.edu/ipm/ipddl/publications/5040e/ [Accessed Jan2018].

Kalidas, P. (2012). Pest problems of oil palm and management strategies for sustainability. Agrotechnology, 11, S001.

Lillesand T. M. (2014). Concepts and Foundations of Remote Sensing Remote Sensing and Image Processing 7thed, Lillesand T M, Kiefer R W &Chipman J W (eds) (United states: WILEY)

Lindgren, C. J. (2012). Biosecurity policy and the use of geospatial predictive tools to address invasive plants: updating the risk analysis toolbox. Risk Analysis: An International Journal, 32(1), 9-15.

Liu, Z., Huang, J., & Tao, R. (2008). Characterizing and Estimating Fungal Disease Severity of Rice Brown Spot with Hyperspectral Reflectance Data. Rice Science, 15(3), 232-242.

Manas J. (2017). Gis Resources: Geospatial Technologies for Sustainable Agriculture Development Issue 3. [Online] Available: [Accessed Jan2018].

Mandal, D., Ghosh, P. P., & Dasgupta, M. (2012). Appropriate precision agriculture with site-specific cropping system management for marginal and small farmers. Plant Sciences Reviews, 2013, 121.

Meeta P. M. (2017). Geo Spatial Technologies for Agriculture [Online] Available: https://geospatialworldforum.org/speaker/SpeakersImages/geo-spatial-technologies-for-agriculture. pdf [Accessed Jan2018].

Misra, H. P. (2008). Bio-efficacy of chlorantraniliprole against shoot and fruit borer of brinjal, Leucinodes orbonalis Guenee. Journal of Insect Science, 24(1):60-64.

Murray A. T. (2017). GIS in regional research In Regional Research Frontiers Vol. 2 ed

Randall J and Peter S Springer International Publishing p169-180 DOI10.1007/978-3-319-50590-9.

Nutter, F. W., & Madden, L. V. (2009). Plant pathogens as biological weapons against agriculture. Beyond Anthrax: The Weaponization of Infectious Diseases, 335-363.

Palaniswami, C., Gopalasundaram, P., & Bhaskaran, A. (2011). Application of GPS and GIS in Sugarcane Agriculture. Sugar Tech, 13, 360-365.

Pandak, P., Bin Ishak, B. I., & Sutha, V. (2014). GIS for rice field management. Geospatial World, 5(1), 40.

Patnaik, H. P. (2000). Flower and fruit infestation by brinjal shoot and fruit borer, Leucinodes orbonalis Guen. damage potential vs. weather. Vegetable Science, 27(1), 82-83.

Ploetz, R., Freeman, S., Konkol, J., Al-Abed, A., Naser, Z., Shalan, K., & Israeli, Y. (2015). Tropical race 4 of Panama disease in the Middle East. Phytoparasitica 43, 283-293.

Pohl, C., Kanniah, K. D., & Loong, C. K. (2016, July). Monitoring oil palm plantations in Malaysia. In 2016 IEEE International Geoscience and Remote Sensing Symposium (IGARSS) (pp. 2556-2559). IEEE.

Prasanna, B. M., Huesing, J. E., Eddy, R., & Peschke, V. M. (2018). Fall Armyworm in Africa: A Guide for Integrated Pest Management, First Edition. Mexico, CDMX: CIMMYT

Qu, Y., & Tao, B. (2014). The constitution of vegetable traceability system in agricultural IOT. Journal of Chemical and Pharmaceutical Research, 6(7), 2580-2583.

Robin, A. H. K., Hossain, M. R., Park, J. I., Kim, H. R., & Nou, I. S. (2017). Glucosinolate profiles in cabbage genotypes influence the preferential feeding of diamondback moth (Plutella xylostella). Frontiers in plant science, 8, 1244.

Sood, K., Singh, S., Rana, R. S., Rana, A., Kalia, V., & Kaushal, A. (2015). Application of GIS in precision agriculture. In Paper presented as lead lecture in national seminar on "Precision farming technologies for high Himalayas (pp. 04-05).

Subbarathnam, G. V., & Butani D. K. (1982). Chemical control of Insect pest complex of brinjal. Entomon, 7, 97-100.

Tayari, E., Jamshid, A. R., & Goodarzi, H. R. (2015). Role of GPS and GIS in precision agriculture. Journal of Scientific Research and Development, 2(3), 157-162.

Thomas, C. S., Skinner, P. W., Fox, A. D., Greer, C. A., & Gubler, W. D. (2002). Utilization of GIS/GPS-based information technology in commercial crop decision making in California, Washington, Oregon, Idaho, and Arizona. Journal of Nematology, 34(3), 200-206.

Thompson, R. B. (1998). Global positioning system: the mathematics of GPS receivers. Mathematics magazine, 71(4), 260-269.

Wójtowicz, M., Wójtowicz, A., & Piekarczyk, J. (2016). Application of remote sensing methods in agriculture. Communications in biometry and crop science, 11(1), 31-50.

Xin, J., Harmon, C. L., Vergot III, P., & Jin, H. (2009). Tracking Pests and Plant Diseases through Space and Time Using Geo-tagged Digital Images. In 7th World Congress on Computers in Agriculture Conference Proceedings, 22-24 June 2009, Reno, Nevada (p. 1). American Society of Agricultural and Biological Engineers.

Yahaya S. M., Musa N. C., Mansor Z., Siham M. N. & a/p Veloo S (2015). Remote Sensing and GIS Web-Based System for Paddy Cultivation Management in Malaysia. Paper presented at 36th Asian conference of remote sensing 2015 Manila, Philippines: ACRS

Zandalinas, S. I., Mittler, R., Balfagón, D., Arbona, V., & Gómez-Cadenas, A. (2018). Plant adaptations to the combination of drought and high temperatures. Physiologia plantarum, 162(1), 2-12.

4

Improvement of Major Cereal Crops for Changing Disease and Insect Pest Scenario in India

Satyendra[1], Riya Raj[2], Sreejaya Ghosh[2] and Prerna Suman[2]

[1]*Department of Plant Breeding and Genetics, Bihar Agricultural University, Sabour, Bhagalpur, Bihar*
[2]*Department of Plant Breeding and Genetics, Bihar Agricultural University Sabour, Bhagalpur, Bihar*

Abstract

Sustainable agricultural production is endangered by several ecological factors, including biotic & abiotic stresses. Biotic stress is the stress as a result of damage done to an organism by other living organisms. The agents of biotic stress are fungi, bacteria, virus, nematodes, insects, weeds & other plants which reduces quality and production of crops. These challenging environmental factors may have adverse effects on future agriculture production. Breeding resistant cultivars is considered as the best strategy to mitigate the effect of biotic stress. It is durable, economical & environment friendly and provide broad-spectrum resistance. Widely cultivated but susceptible genotypes can be improved by introgression of resistance genes from donor genotype into the background of recurrent genotype by different breeding strategies. In modern agriculture, conventional crop-breeding strategies alone are inadequate for achieving the increasing population's food demand on a sustainable basis. The advancement of molecular genetics and related technologies are promising tools for the selection of new crop species with resistance genes. Transfer of several genes or quantitative trait loci (QTL) with potential characteristics into a single widely cultivated genotype is possible through the process of marker assisted selection (MAS). Gene pyramiding through MAS have accelerated the development of durable resistant/tolerant lines with high accuracy in the shortest period of time. With marker assisted backcross breeding (MABB), a major gene can be precisely targeted, the genome of the recurrent parent can be recovered fast, and linkage drag can be reduced. Genomic selection (GS) allows selection for multiple traits directly in the genome. These techniques help in identifying plants with two or more genes by employing gene-linked markers without the need for repeated testing in each generation during the development of pyramided lines and hence shorten the time of breeding program.

Keywords: *cereals, climate change, diseases, insect pests, maize, rice, wheat*

Introduction

The three major cereals in India, in-terms of area under production and yield are rice (*Oryza sativa* L.), bread wheat (*Triticum aestivum* L.) and maize (*Zea mays* L.). Worldwide, cereals (i.e., grain crops rich in carbohydrates) are the most extensively cultivated and consumed crops. When all cereals are combined, they supply around 56% and 50% of the world's protein and calorie needs, respectively. The most widely produced and consumed cereals worldwide are maize, wheat, rice, barley, and sorghum. They are extensively cultivated throughout the world, from the tropics to the temperate zones. Huge yield shortfalls persist despite attempts made by governments, research groups, and academic institutions to boost the productivity of these essential crops. Abiotic stresses like heat and drought, as well as biotic ones like pests and diseases brought on by the climate, are generally recognized as the main factors limiting the productivity of most cereal crops. Given the major role that cereals play in the security of the world's food supply and nutrition, improvements in the productivity of cereal crops are mandatory for meeting the demands of the population.

Although these cereals play a significant role in the security of the world's food supply and nutrition, biotic (such as pests and diseases) and abiotic (such as heat and drought) stresses continuously reduce their production (Kosina *et al.*, 2007). In the field and during post-harvest storage, they lower crop production and quality (Zaidi *et al.*, 2003). However, crop species and variety, the severity and duration of crop stress, and the developmental period at which the stress occurs all affect how much these restrictions affect cereal production and yield quality (Tadese *et al.*, 2019). Biotic stresses are the ones that significantly affect these cereals' yield losses among other stresses. Cereal production should rise to meet the demand for food, feed, and industry to ensure food and nutrition security for the present and the future (Prasanna *et al.*, 2013). By lessening the effect of biotic stress on cereals, the efficiency of the systems used to produce grains can be increased. Furthermore, it is well established that genetic crop improvement, utilizing both traditional and molecular breeding methods, is a crucial crop adaptation approach for the future socio-climatic conditions that are anticipated.

Insect pests and diseases are the major biotic stresses that limit rice production. The environment in which rice is farmed influences the frequency of diseases and insect pests. Depending on the prevalent determinants of pest abundance in a given year or season, the level of loss resulting from various biotic stresses varies greatly. Rice is attacked by several insect pests during different stages of the crop. At the initial stage, termites, ants, grasshoppers, thrips, and swarming caterpillar; at the early tillering stage, again thrips, case worm, gall midge, rice hispa, leaf folders, swarming caterpillar, stem borer and cutworm are the major insect pests while at the late tillering stage leaf folder, brown plant hopper, white backed plant hopper, yellow stem borer are the major insects and at the panicle initiation stage, stem borers, leaf folders; at flowering stage gundhi bug and at maturity ear-cutting caterpillar are the major pests of concern. Ecosystem-wise, the occurrence of different insect pests is termite, red ants, grasshoppers, gundhi bug, mealy bug in upland; in irrigated/rainfed favourable shallow lowland thrips, gall midge, swarming caterpillar, rice hispa, cutworm, leaf folders, mealy bug, yellow stem borer, brown plant hopper, white backed plant hopper, gundhi bug and cut worm whereas in rainfed lowland cutworm and yellow stem borer are the major insect pests recorded. Thus, irrigated ecosystem harbours maximum number of insect pests.

Table 1: Insect-pest scenario of rice in India

Major Insect Pests of Rice						
Common Name	**Scientific Name**	**Extent of damage**	**Occurrence**	**Prevalent races**	**Resistance genes/ QTLs**	**Donor parent**
Yellow stem borer	*Scirpophaga incertulus*	Severe Moderate to severe Low to moderate 20-60%	Throughout the country		*Cry* genes	*O. longistaminat a*
Gall midge	*Orseolia oryzae*	Severe Moderate to severe 10-20 %	Throughout the country	Biotype 1-7 (Devi *et al.*, 2021)	*Gm1, Gm3, Gm5, Gm6* and *Gm8* (Devi *et al.*, 2021)	PTB10, SIAM29, Duokang#1, Abhaya, Surekha, Samridhi, Ruchi, Asha
Brown planthopper	*Nilaparvata lugens*	10-50 %	Throughout the country	Biotype 1-4 (Jena *et al.*, 2010)	*Bph1, Bph33(t), bph2, Bph3, Bph14 and Bph15* (Jena *et al.*, 2010)	IR26, IR36, Rathu Heenati, Swarnalata, Pokkali, PTB33
Leaf folder	*Cnaphalocrocis medinalis*	Moderate to severe Moderate to most Severe 5-25 %	Throughout the country		*Sck* and *crylac (Rao et al., 2010)*	Cauveri, Akashi, TKM 6, TKM 12, ADT 46, TPS 2, TPS 3, ADT 44, PY 4, Kairali, Ahalya,
Green leaf Hopper	*Nephotettix virescens*	Moderate to severe Low to moderate	Assam, Bihar, Madhya Pradesh, Andhra Pradesh, Tamil Nadu, Uttar Pradesh and West Bengal		*GRH6, Glh1, Glh2, Glh3, Glh5, Glh6, Glh7, Glh9, Glh10, Glh11, Glh12, Glh13, Glh14, glh4, glh8* and *glh10* (Phi *et al.*, 2019)	Pe-bi-hun, IR24, Singwang, Lepedumai, DV85, SML17, TAPL 796, Ptb 18 and IR36

Major Insect Pests of Rice						
Common Name	**Scientific Name**	**Extent of damage**	**Occurrence**	**Prevalent races**	**Resistance genes/ QTLs**	**Donor parent**
Rice Thrips	*Stenchaetothrip s biformis*	5-20%	Kerala, Tamil Nadu, Upland areas in Odisha, Andhra Pradesh, Madhya Pradesh, Punjab, Haryana, Assam			
Minor Insect Pests of Rice						
White backed plant hopper	*Sogatella furcifera*	Moderate to severe Light to moderate 1-5 %	Madhya Pradesh, Punjab, Uttar Pradesh, Bihar, Gujarat, Haryana,		*Wbph1, Wbph2, F3H, Wbph3, Wbph4, Bph38, Wbph7* and *Wbph8* (Padmavathi *et al*., 2007)	ADR 52, Podiwi A8, ARC 6650, ARC 5984, Velluthecherr a, and MO I
			Himachal Pradesh and Odisha			
Rice Hispa	*Dicladispa armigera*	Moderate to severe Light to moderate	Bihar, West Bengal, Assam, Odisha, Meghalaya, Mizoram, Tripura, Punjab, Himachal Pradesh, Uttar Pradesh and Uttarakhand			
Gundhi bug	*Leptocorisa acuta*	Light to moderate 5-60 %	Throughout the country			
Swarming caterpillar	*Spodoptera mauritia*	Light to moderate	Low-lying areas in Bihar, Odisha, West Bengal, Assam, Jharkhand, Punjab and Chhattisgarh			

Termite	*Odontotermes obesus*		Rainfed upland areas and irrigated rice-wheat system			
Sporadic Insect Pests						
Case worm	*Nymphula depunctalis*		Kerala, Odisha, Tamil Nadu and Low-lying water-logged areas in eastern India			
Cutworm	*Mythimna separata*	5 %	Uttar Pradesh, Coastal areas, Haryana and Punjab			
Mealy bug	*Brevennia rehi*	5-10 %	Upland areas in Uttar Pradesh, Bihar, West Bengal, Odisha, Madhya Pradesh, Tamil Nadu, Kerala, Pondicherry and Karnataka			
Root weevil	*Echinocnemus oryzae*		Bihar, Haryana, Punjab and Tamil Nadu			

Source: Prakash *et al.* (2014)

(A) (B) (C)

Fig. 1: A. Rice stem borer, **B.** Rice gall midge, **C.** Rice brown plant hopper

Rice production and farmers' lives can be severely impacted by rice diseases. Covering one-fourth of the total planted area, it is one of the most significant food crops in India. For more than half of the world's population, rice is the main diet. In terms of rice output worldwide, India comes in second to China. The total amount of rice produced in 2022–2023 is 125 million tons. With an average output of around 4.1 tonnes/ha, 45.5 million hectares will be under rice cultivation in 2022–2023. In India, the Kharif season is when paddy is mostly farmed. It thrives in hot, humid, tropical and subtropical regions. The severity of the diseases increased with large regions under monoculture and intensive rice farming; numerous main diseases got even more severe, and minor diseases manifested as epidemics. The predominant weather conditions and agricultural methods in the region mostly determine the severity. Before the introduction of exotic varieties, diseases such as sheath blight, sheath rot, and rice tungro disease were either insignificant or unheard of. Blast, bacterial leaf blight, sheath blight, brown leaf spot, and false smut have been identified as the major rice diseases as of late. Of these, the false smut and the sheath blight gained significance only following the Green Revolution. (Chander *et al.*, 2003).

Table 2: Diseases scenario of rice in India

Major Diseases of Rice						
Common Name	**Scientific Name**	**Extent of damage**	**Distribution**	**Prevalent races**	**Resistance genes/ QTLs**	**Donor parent**
Rice Blast	*Magnaporthe grisea*	70-80%	Himachal Pradesh, Haryana, Karnataka, Andhra Pradesh, Tamil Nadu, Kerala and upland parts of India	ZD1, ZD3, ZD5, and ZE1	*Pi2, Pi9, Piz, Pita, Pita2, Pib, Piz(t), Pi54, Pi33, Pi36, Pi21, Pi50, Pi65, qBR4-2, qBR12-1* (Yadav *et al*., 2019, Gavhane *et al*., 2019)	Tetep, Modan, Digu, K59, Toride 1, Kahei, K60, Katy, Q61, Gumei2, Gumei 4
Bacterial Leaf Blight	*Xanthomonas oryzae*	20-80%	Punjab, Rajasthan Haryana, Western Uttarpradesh, Uttarkhand plains, Pallakad district of Kerala and parts of Andra Pradesh	A4 to A9 (Tekete *et al*., 2020)	*Xa1, Xa3, Xa5, Xa13, Xa21, Xa26, Xa18, Xa34, Xa25, Xa41, Xa46(t)* (Tekete *et al*., 2020, Chen *et al*., 2020, Ji *et al*., 2018)	IRBB1, IRBB2, Kogyoku, Chukogu-45, IRBB3, IRBB4, DV85, IRBB5
Sheath Rot	*Sarocladium oryzae*	20-85%	Jammu & Kashmir		*qShB4-1, qShB9* (Hittalmani *et al*., 2002)	Swarna, Dhalaheera, JGL 3855 and Kattanur
Rice Brown Spot	*Helminthosporium oryzae*	50 %	Throughout the country	Iga-2	*qBS2, qBS9* and *qBS 11, bsr1* (Matsumoto *et al*., 2021)	Tadukan, CH45, Tetep
False Smut	*Ustilaginoidea virens*	2.8-80%	Uttar Pradesh, Punjab, Bihar		*qFsr8–1, qFsr2–1, qFsr4–1, qFsr8–2, qFsr11–1, qFsr10* and *qFsr12* (Andargie *et al*., 2018)	IR28, IR36, Ranjeet, Nongxiang 21, Luxiang 90-1 and Shuangkang 7701

Major Diseases of Rice						
Common Name	**Scientific Name**	**Extent of damage**	**Distribution**	**Prevalent races**	**Resistance genes/ QTLs**	**Donor parent**
Sheath Blight	*Rhizoctonia solani*	20-69 %	Uttar Pradesh, Uttarakhand, Bihar, West Bengal, Haryana, Odisha, Chhattisgarh, Tamil Nadu, Kerala, Karnataka, Andhra Pradesh, Jammu and Kashmir, Madhya Pradesh, Assam, Tripura and Manipur	AG1-IA (Zheng *et al.*, 2013)	*qShB9-2, qAB9-TQ, qSB7Tq, qSB9Tq, qSB11Le* and *qSBR11-1* (Li *et al.*, 2022, Molla *et al.*, 2020)	Bharti, CR-1014, Nalini, Pankaj, Ratna, Tetep, Jasmine 85, Tequing and MCR010277
Tungro Disease	*Rice Tungro virus*		Assam, Bihar, Karnataka, Kera!a, Manipur, Orissa, Tamil Nadu, Tripura, Uttar Pradesh, and West Bengal	*RTSV and RTBV*	*CP3, eIF4G, tsv1* (Hore *et al.*, 2022)	Matatag 1, Basmati 370, Habiganj DW8, Kataribhog, MR 81, Pankhari 203, Utri Merah, Utri Rajapan and Y 1036
Minor Diseases of Rice						
Leaf streak	*Xanthomonas oryzae pv. oryzicola*	8-32%	Haryana, Rajasthan, Uttarakhand, and Uttar Pradesh	Thai *Xoc* isolate, SP7-5 and 2NY2-2	*Xo2, Xa5, Xo1* and *bls1* (Xie *et al.* 2014)	IR62266, MNTK75, DV85
Foot rot / Bakanae	*Gibberella fujikuroi*	70%	Eastern and north-eastern states of India	FGSC 7600 (MATA-1), FGSC 7603 (MATA-2), FGSC 7611 (MATB-1), FGSC 7610 (MATB-2) (Leslie *et al.*, 2004)	*d1, d29, sd6* or *sdq(t))* *qBK1.1, qFfR1 qBK1.2, qBK1.3, qBK1z* (Lee *et al.*, 2021, Fiyaz *et al.*, 2016)	Quingxi, Longjiao 86074-6, Zupei 7, Dongrong 84-21, G-6, Sui 89-17, C-101A51, Athad Apunu, Chandana, and Panchami

(A) (B) (C)

(D) (E)

Fig. 2: A. Rice blast, **B.** Sheath blight, **C.** Brown spot, **D.** False Smut, **E.** Bacterial leaf blight

Table 3: Insect pest scenario of wheat in India

Insect pest	Scientific name	Extant of damage	Distribution	Prevalent races	Resistant genes	Donor parent
Termite	*Odontotermes obesus* (Rambur), *Microtermes obesi* Holmgren (Satyagopal *et al*.,2014)	20-40% damage (Sharma *et al*., 2004)	Himachal Pradesh, Jammu and Kashmir, Punjab, Haryana, Uttar Pradesh, Rajasthan, Madhya Pradesh, Bihar, Maharashtra and Karnataka (Sharma *et al*., 2004)	*Odontotermes bellahunisensis* Holmgren & Holmgren, *O. obesus* (Rambur), *O. redemanni* (Wasmann), *Microtermes obesi* Holmgren, M. *mycophagus Desneux*, *Heterotermes indicola* (Wasmann) (Mahapatro *et al*., 2013)	-	-
Green bug	*Schizaphis graminum*(Satyagopal *et al*.,2014)	have been reported up to 36 %.	Delhi, Haryana, Himachal Pradesh, Jammu and Kashmir, Madhya Pradesh, Punjab, Rajasthan, Uttar Pradesh and West Bengal. (Sharma *et al*., 2004)	Multiple biotypes (≥ 10) Biotypes 1 and 2 are prevalent	Gby, Gbz, Gb3 (Boyko *et al*., 2004), (Weng *et al*., 2002) (Boina *et al*., 2005)	Gby gene in the germplasm 'Sando's 4040' (Boyko *et al*., 2004) Gbz gene from Ae tauschii (Zhu *et al*., 2004).
Armyworm	*Mythimna separata* (Walker) (Satyagopal *et al*.,2014)		Prevalent in U.P, Bihar, Rajasthan, and Punjab. (Sharma *et al*., 2004)			
Hessian fly	*Mayetiola destructor* (Satyagopal *et al*.,2014)	5-10 % damage	Not prevalent in india.	Hf biotype L and Hf biotype J, Hf biotype D (Tadesse al., 2013), (Maas al., 1987)	H1 to H28(Ratcliffee t al., 2013)	8 genes from common wheat, 13 genes from durum wheat, 5 from *T. tauschii*, and 2 from rye.(steuart al., 2008)

Insect pest	Scientific name	Extant of damage	Distribution	Prevalent races	Resistant genes	Donor parent
Pink stem borer	*Sesamia inferens* (Walker) (Satyagopal et al.,2014)	“dead heart”, “white ears” Reduced yield by more than 11 per cent in India (Sharma *et al*., 2004)	North-Western plains of India			
Shoot fly	*Atherigona naqvii* Steyskal, *A. oryzae* Mall (Satyagopal *et al*.,2014)	Cause 20 to 37 per cent yield loss (Jambai *et al*., 2021)	m Rajasthan, Gujarat, Madhya Pradesh, Uttar Pradesh, Punjab and Haryana (Sharma *et al*., 2004)			
Ghujhia weevil	*Tanymecus indicus* Faust (Satyagopal *et al*.,2014)	Cut the germinating seedling at ground levels and often crop has to be re- sown (Sharma *et al*., 2004)	Widely distributed in the Indian Sub- continent mostly Uttar Pradesh, Bihar and Punjab (Sharma *et al*., 2004)			

Table 4: Diseases scenario of wheat in India

Diseases	Scientific name	Extant of damage	Distribution	Prevalent races	Resistant genes	Donor parent
Stem rust	*Puccinia graminis f. sp. tritici*	can cause up to 100% crop loss	Hills, foothills and plains of north western India and southern hills zone (Nilgiri hills of Tamilnadu).	*Ug99*,40-2,40-3,184-1,15-1 (Bhardwaj *et al.*,2019)	Sr2, Sr5, Sr13, Sr22, Sr24, Sr25, Sr26, Sr27, Sr28, Sr32, Sr33, Sr35, Sr36, Sr37, Sr39, Sr40, Sr44 and Sr Tmp (Anonymous 1)	Sr14 from *T. turgidum* Sr24/Lr24, Sr25/Lr19 and Sr 26 from *Agropyron elongatum* Sr22 from *T. monococcum* Sr32, Sr33/Lr21 from *Aegilops speltoides* Sr36/Pm6 from *Triticum timopheevii* Sr38/Lr37/Yr17 from *Aegilops ventricose* Sr31/Lr26/Yr9/Pm8, Sr27 from *Secale cereale* (Anonymous 1)
Leaf rust	*Puccinia triticina* (formally *Puccinia recondita*)	45-50% crop loss	Hills, foothills and plains of north western India and southern hills zone (Nilgiri hills of Tamilnadu).	77-6,77-7,77-8,77-9,77-10,77-11 (Bhardwaj *et al.*,2019)	Lr9, *Lr19*, *Lr*2 4, *Lr*25, *Lr*29, *Lr*32, *Lr*39, *Lr* 45, and *Lr*47 (Ano nymous1)	
Stripe rust	*Puccinia striiformis*	can cause up to 100% crop loss	Hills, foothills and plains of north western India and southern hills zone (Nilgiri hills of Tamilnadu).	78S84, 46S117, 110S84, 110S119, 110S247, 238S119 (Bhardwaj *et al.*,2019)	Yr*5*, Yr*10*, Yr *15*, Yr*Sp*, *and* Yr*Sk* (Anonymous 1)	
Powdery mildew	*Blumeria graminis f. sp. tritici*	yield losses as high as 48%	North Western Plain zone covering Uttarakhand, U.P., Punjab, Haryana, Rajasthan and Delhi		Pm3c, Pm5 and Pm8 Pm 1, Pm 2, Pm 3b, Pm 4a, Pm 13 and Pm 20 are rare and these can be exploited. (Anonymous 2)	Pm6 from *Triticum timopheevii* Pm8 from *Secale cereale* (Anonymous 1) PmTb7A.1 and PmTb7A.2 from *T. boeoticum* accession pau5088 (Anonymous 2)

Diseases	Scientific name	Extant of damage	Distribution	Prevalent races	Resistant genes	Donor parent
Loose smut	*Ustilago tritici*	2-3 % of the total yield	All wheat growing regions of India, particularly in Punjab, Haryana, U.P. and certain regions of M.P.	T1, T10, T11, T32 T33, T26, T14, T3, and T4 (Menzies *et al.*, 2003)	Ut6, Ut11 (Thambugala *et al.*, 2020), (Kassa *et al.*, 2014)	In long arm of 5B chromosome (Kassa *et al.*, 2014)
Blast	*Magnaporthe oryzae* triticum	Average yield loss was estimated at 25-30 percent	in districts close to West Bengal borders	pathotype (MoT) (Singh *et al.*, 2021)		
Fusarium head blight	*Fusarium graminearum*	10-70% yield loss during epidemic years	Humid and semi-humid regions and foot hills of Punjab, Himachal Pradesh, Uttarakhand and hilly areas in Tamil Nadu.		WFhb1(Unpub lished (Paudel *et al.*, 2020)	

Table 5: Insect pest scenario of maize in India

Common name	**Scientific name**	**Distribution**	**Resistance genes/ QTLs**	**Donor parents**
Maize stem borer	*Chilo partellus*	This pest is distributed throughout maize growing regions of India in kharif and spring season.	CRY1C, CRY2A	MH63
Pink stem borer	*Sesamia inferens*	Andhra Pradesh, Assam, Bihar, Delhi, Karnataka, Madhya Pradesh, Maharashtra, Orissa, Punjab, Tamil Nadu, Uttar Pradesh and West Bengal in winter season, and also in spring in Haryana and Punjab.		
Shoot fly	*Atherigona* spp.	Delhi, Haryana, Punjab, Rajasthan, Uttar Pradesh, Andhra Pradesh, Bihar and Tamil Nadu.	qDH9	CM143 *CM144
Cutworm	*Agrotis ipsilon*	Bihar, Haryana, Madhya Pradesh, Himachal Pradesh, Orissa, Punjab, Rajasthan, Uttar Pradesh and West Bengal		
Army worm	*Mythimna separata*	Commonly occurs in Andhra Pradesh, Assam, Bihar, Delhi, Gujarat, Kerala, Karnataka, Madhya Pradesh, Maharashtra, Manipur, Orissa, Punjab, Tamil Nadu, Rajasthan, Uttar Pradesh and West Bengal.	Cry1F	TC3507
Aphid	*Rhopalosiphum maidis*	This pest is distributed throughout India	Sod2, sod3.4, sod9, sodB	*Zea mays* L. tasty sweet(susceptible) and *ambrozja*(relatively resistant)
Thrips	*Frankliniella occidentalis*	This pest is distributed throughout India		
Shoot Bug	*Peregrinus maidis*	Karnataka, Tamil Nadu, Andhra Pradesh and Madhya Pradesh		
Grass hopper	*Hieroglyphus nigrorepletus*	It is a sporadic pest distributed all over India but more prevalent in Rajasthan.		
Leaf roller or Maize web worm	*Marasmia trapezalis*	Maharashtra, Karnataka, Andhra Pradesh, Delhi and Uttar Pradesh.		

Source: Kumar *et al*. (2018)

Table 6: Disease scenario of maize in India

Common Name	Scientific Name	Extent of damage	Distribution	Prevalent races	Resistance genes/ QTLs	Donor parent
Northern leaf blight	*Exserohilum turcicum*	13-50%	Karnataka, Maharashtra, Andhra Pradesh, Himachal Pradesh, Meghalaya, Tripura, Assam, Sikkim, Bihar and Jharkhand	*CCR 1* (Hurni *et al*., 2015)	*Htn1, Hm1, remorin, Ht2/Ht3, GST* (Wisser *et al*., 2011, Jamann *et al*., 2016)	PEPTILLA, *Tripsacum floridanum*
Southern leaf blight	*Helminthosporiu m maydis*	15-46%	Jammu & Kashmir, Himachal Pradesh, Rajasthan, Andhra Pradesh, Uttar Pradesh, Karnataka and Tamil Nadu	Race O, Race T, Race C	*ZmFUT1, MYBR92, rhm1, qMdr9.02* (Chen *et al*., 2023)	*Zea mays*
Gray leaf spot	*Cercospora zeae-maydis*	5-30%	Throughout the country		*Qgls8, qRgls1.06* (Sun *et al*., 2021)	*Zea parviglumis*
Common rust	*Puccinia sorghi*	18-49%	Karnataka, Andhra Pradesh, Bihar, Madhya Pradesh, Maharashtra, Rajasthan, Mizoram and West Bengal		*Rp1-D* (Collins *et al*., 1999)	PIC20
Head smut	*Sphacelotheca reiliana*	Up to 30%	Karnataka & Tamil Nadu		*ZmWAK, qRfg1, qRfg2* (*Zuo et al., 2015*)	Jidan558, Ji1037
Anthracnose stalk rot	*Colletotrichum graminicola*	10-42%	Andhra Pradesh, Karnataka and Rajasthan		*Rcg1* (Frey, 2023)	INBRED LINE- MP305
Sugarcane mosaic virus disease	*Potyvirus*	20-30%	Throughout the country		*ZmTrxh, ZmABP1* (Liu *et al*., 2017)	
Stewart's wilt	*Erwinia stewartii*	40-100%			*pan1* (Jamann *et al*., 2016)	

Breeding Strategies for Biotic Stress Resistance

Biotic stress is the stress as a result of damage done to an organism by other living organisms. The agents of biotic stress are fungi, bacteria, virus, nematodes, insects, weeds & other plants that reduces the quality and production of crops. These challenging environmental factors may have adverse effects on future agriculture production.

Why Resistance is Needed?

Biotic stress significantly limits global crop production. Pesticides are costly and need multiple applications. Chemical control is not effective. Accumulation of toxic chemicals in the crops and they are hazardous to human health and for the environment. Pest resistance to chemical pesticides. Nowadays, resistant cultivars are currently seen as the best strategy as it is durable, economical, broad-spectrum and environment-friendly.

Genetics of Disease Resistance- It was first categorized by Van der Plank (1963,1968) into two-

1. Vertical resistance (qualitative)
2. Horizontal resistance (quantitative)

Table 7: Features of qualitative and quantitative resistance

Category	Vertical resistance	Horizontal resistance
Synonyms	Qualitative, Differential	Quantitative, uniform
Pathogen specificity	Race-specific	Race non-specific
Degree of resistance	Complete	Partial
Mechanism	Hypersensitivity	Diverse
Plant growth stage	All stage (seedling resistance)	Different (adult plant resistance)
Assessment	Infection type	Disease severity
Durability	Low	High
Inheritance	Monogenic	Polygenic
Gene effect	Major	Minor

Breeding Strategies

There are 2 types of breeding strategy-

1. Conventional - selection is done using natural processes
2. Advanced – selection is done with the use of molecular tools

Conventional Breeding Strategies

1. Backcrossing
2. Pedigree breeding
3. Recurrent selection

Backcrossing- It involves crossing F1 with one of the parental lines, followed by selection for the desired characteristic. The backcrossing method is also used for resistance gene pyramiding. A donor genotype having a particular desirable gene (red) is crossed with an elite genotype (recipient) & F_1 is repeatedly crossed with the recipient parent (elite). Backcrossing includes selection for the interest gene and recovery of an increased proportion of the recipient genome.

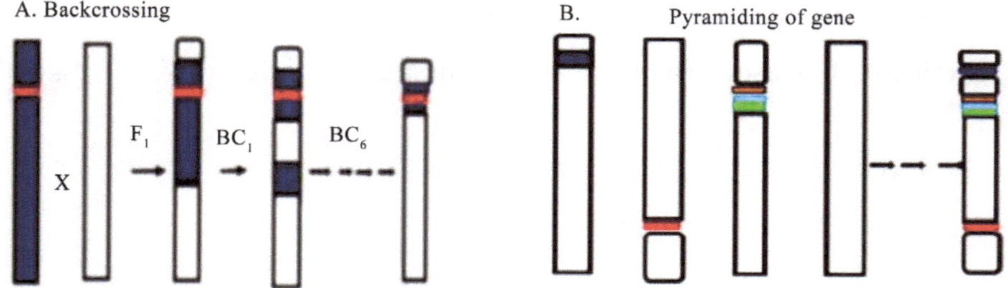

Fig. 3: Back-crossing and pyramiding of genes

Pedigree breeding- It is a method of genetic improvement of self-pollinated species in which superior genotypes are selected from segregating generations and proper records of the ancestry of selected plants are maintained at each stage of selection. Two individuals with suitable and compatible phenotypes are crossed; the F_1 offspring is self-pollinated to get new, improved variations of the genotype.

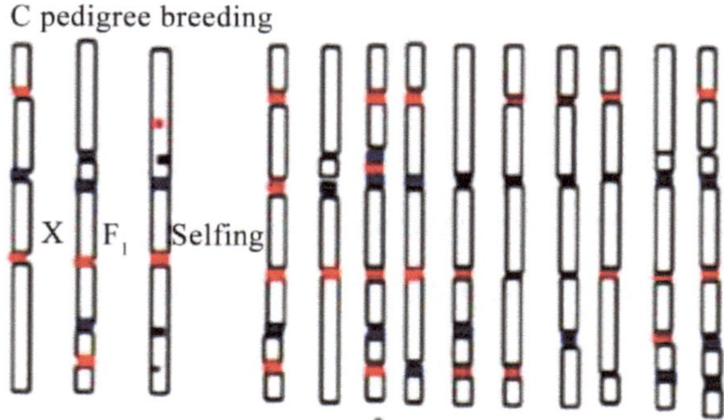

Fig. 4: Pedigree breeding

Recurrent selection- It is an efficient and modified form of progeny selection, where selection for some specific trait(s) is conducted within consecutive segregating progeny generations on the basis of phenotypic characteristics. An individual population having two traits each of which is influenced by two major favorable QTLs. Intermating & selection increases the frequencies of favorable alleles at each locus. Here, no individual in the initial population had all desirable alleles, but after recurrent selection, half the population possessed the desired genotype.

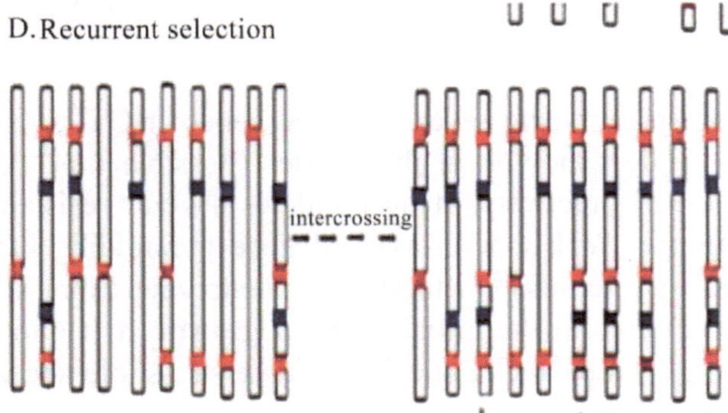

Fig. 5: Recurrent selection

Why Advanced Breeding is needed in the present time?

Advanced Breeding is precise, sophisticated, and rapid breeding through molecular markers. Easy accessibility of polymorphic markers makes advanced breeding successful. There are linkage maps, and QTLs for different quantitative and qualitative traits. It is not affected by environmental factors. It allows breeders to identify and select desirable plants at seedling stage. It precisely identifies genotypes carrying the stacked desired genes.

Advanced breeding strategies

1. Marker-Assisted Selection
   - Marker-Assisted Gene Pyramiding
   - Marker-Assisted Backcross Breeding
   - Marker-Assisted Recurrent Selection
2. Double Haploid breeding
3. Genomic selection
4. Speed Breeding

Marker-assisted Selection

MAS is an Indirect selection of a trait of interest by using molecular marker. It can be used to assist selection of parents & increase the effectiveness of backcross breeding for biotic & abiotic stress resistance. Currently SSRs (second-generation markers) are widely used markers in MAS due to the easy availability and comparatively cheaper than others and they require a comparatively simple technique with a higher polymorphism rate. The success of MAS depends on several factors, including the number of target genes to be transferred and the distance between the flanking markers and the target gene.

Marker-assisted Gene Pyramiding

It is a useful technique for transferring several desired genes or QTLs from different parents into a single genotype in the shortest possible time (2-3 generations) with 99.2% RPG recovery. It accumulates several resistance genes into a single genotype. With marker-assisted gene pyramiding QTL allele-linked markers are easy to select with similar phenotypic expression and desirable plants are selected at the initial stage of growth. It is Quick, proficient, economical, and a simple technique for multiple stress tolerances.

Pyramiding genes/ QTLs involves two steps: Assembling all target genes in a single genotype known as root genotype by multiple crossings and Fixation of the target genes in a single, homozygous genotype.

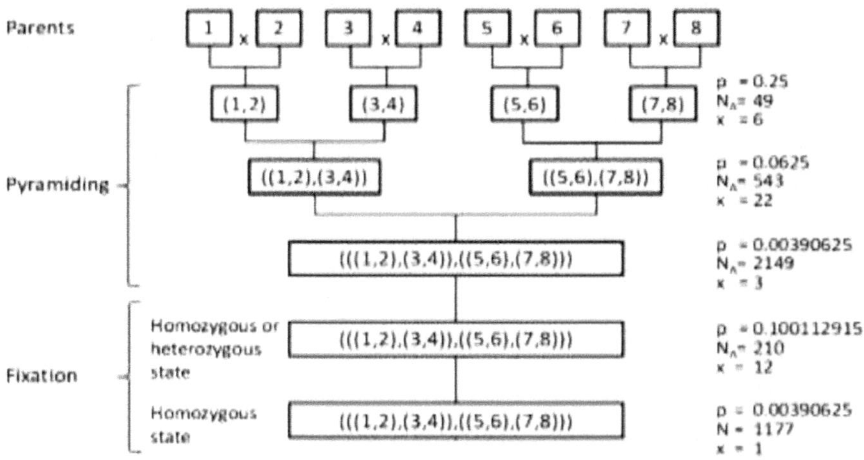

Fig 6: Marker-assisted gene pyramiding

Table 8: Marker-assisted gene pyramiding in crops for biotic stress resistance

Crop	Trait	Pyramided genes
Rice	Bacterial blight resistance	xa5? xa 13, Xa21
	Bacterial blight resistance	Xa4. xa5, xa13 and Xa21
	Blast resistance	Pi1T Pi2 and Pi33
Wheat	Leaf rust resistance	Lr41, Lr42 and Lr43
	Powdery mildew resistance	Pm2+Pm4a. Pm2+Pm21. Pm4a+Pm21)
	Stripe rust resistance	YrSand Yr15
Barley	Barley Yellow Mosaic Virus resistance	rym4, rym5, rym9 and rym11
Potato	Late blight resistance	Rpi-mcd1 and Rpi-b&r
Soybean	Soybean mosaic virus (SMV}-resistance	RsvA t Rsv3t and

Marker-assisted Gene Pyramiding Strategies

There are 3 different backcrossing strategies for gene pyramiding

1. Separate backcross programs
2. Single backcross: Symmetrical mating
3. Single backcross: Tandem mating

Separate backcross programs- In the first approach, each DP is used in a separate backcross program with the RP to recover the target gene from each DP in the genetic background of RP either in heterozygous or homozygous state. These derived lines of RP are then crossed together to produce a complex hybrid. Finally, the pyramided version of RP having all the target genes is recovered from this hybrid by selfing coupled with selection.

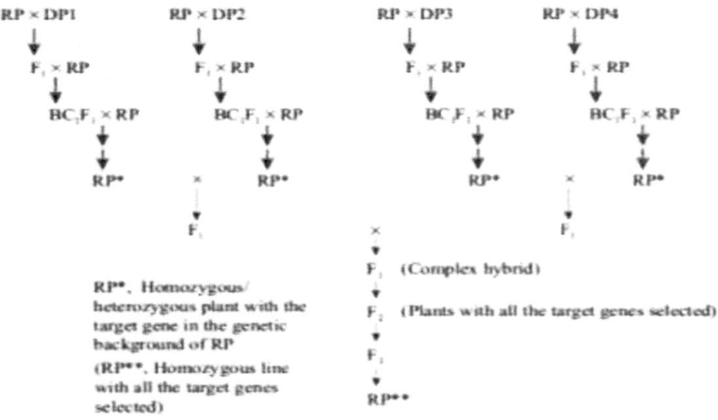

Fig. 7: Separate back-crossing method

Single Backcross: Symmetrical Mating

In the second approach, all the DPs are ordered into a single backcross program according to a suitable mating scheme. In the case of symmetrical mating scheme, the F1s from the crosses between different DPs and the RP are crossed in pairs to

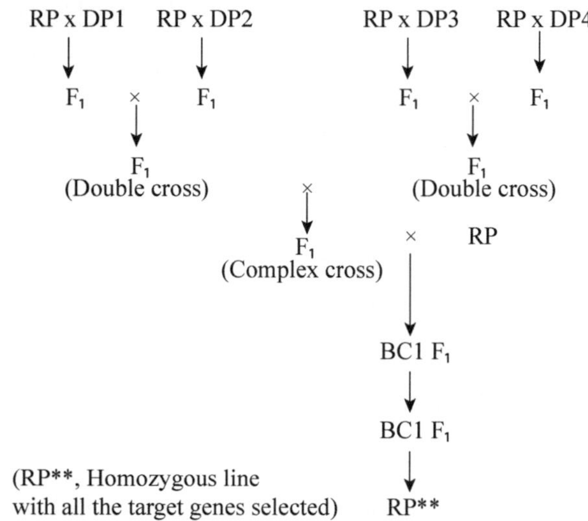

Fig. 8: Single back-cross: symmetrical method

Single backcross: Tandem Mating

The recurrent parent is first crossed with one of the DPs. The F1 from this cross is now mated to the second DP and so on till all the DPs are mated in succession to produce a complex hybrid. The complex hybrid obtained from either scheme is used in a backcross program with the RP to recover the pyramided version of RP.

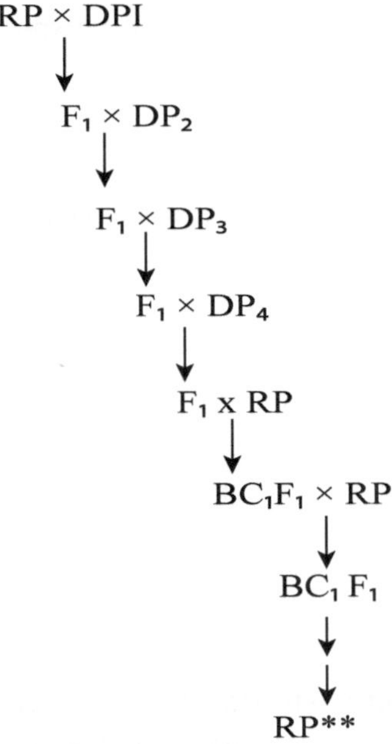

Fig. 8: Single back-cross: Tendem method

Marker-Assisted Backcross Breeding (MABB)

MABB is simple form of MAS targeting one or more genes or QTLs transferred from one donor parent into another superior cultivar or genotype to improve a targeted trait. MABB is advantageous when recessive allele has to backcrossed & the target gene is expressed at later development stage. It is also used for introgression of several genes/QTLs. Contrary to conventional backcrossing, MABB depends on the alleles of a marker linked with desirable genes or QTLs instead of phenotypic performance. Recurrent parent is crossed with donor parent & the F_1 is backcrossed to the recurrent parent up to 3 generations. Then individuals are identified through MAS & desired individuals are advanced to next generation.

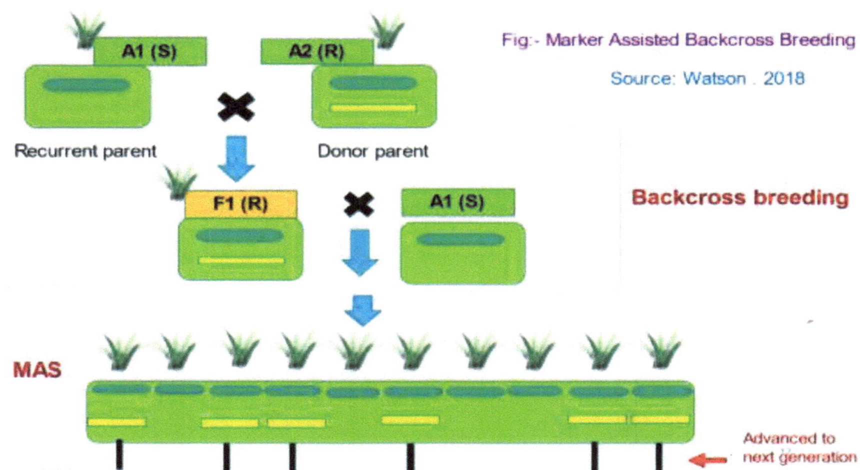

Fig. 8: Marker assisted back-cross method

Three levels of Selection are Done Here

The first level of selection is called **foreground selection** where progenies carrying the desired gene would be selected using markers linked to the target gene or trait of interest (for heterozygous genotype). The second level of selection is **recombination selection** where homozygous alleles of the recurrent parent will be selected using tightly linked markers flanking the targ *et al* leles (done on the carrier chromosome using flanking markers). It reduces linkage drag which is difficult in conventional backcrossing. The third level of selection is called the **background selection** which involves the selection of individuals carrying homozygous alleles of the recurrent parent at several unlinked marker loci covering the entire genome. Background selection is selection towards the recurrent parent which accelerates the recovery of the recurrent parent genome and hence the use of MAS in backcross breeding reduces the time required to recover the recurrent genome by two to four backcrosses.

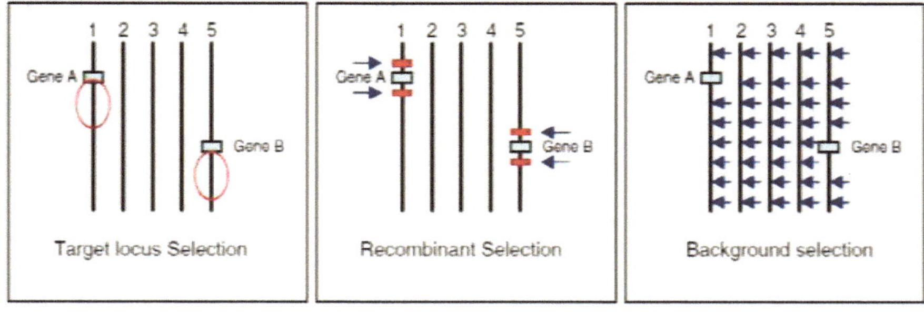

Fig. 10: Level of selection in MABB

Marker-Assisted Recurrent Selection

MARS is also known as Marker-assisted Forward Breeding. The basic procedure of MARS comprises some steps, such as selection of the parent lines from a similar and non-similar population. The best population is selected for recombination & F1s are generated. F_1 individuals are allowed for selfing to produce F_2 progenies using the single-seed decent method. F_2 individuals are then advanced to F_3, F_4, and F_5. Genotyping is carried out on individual progenies and the best seeds are selected and evaluated in multi-location phenotyping. (QTL) analysis is done and modeling is conducted prior to selection of QTLs for recombination. The six best genotypes per offspring of F_3 are selected (A–F). The first recombination cycle generated F_1 individuals, which are selfed to produce; F_2 and F_2 are advanced to F_3 and F_4 for multi-location phenotyping. Large plant numbers are preferred to rely on the accuracy of QTL mapping. Further, QTL can be evaluated after genotyping & phenotyping analysis for the selection of markers and suitable alleles.

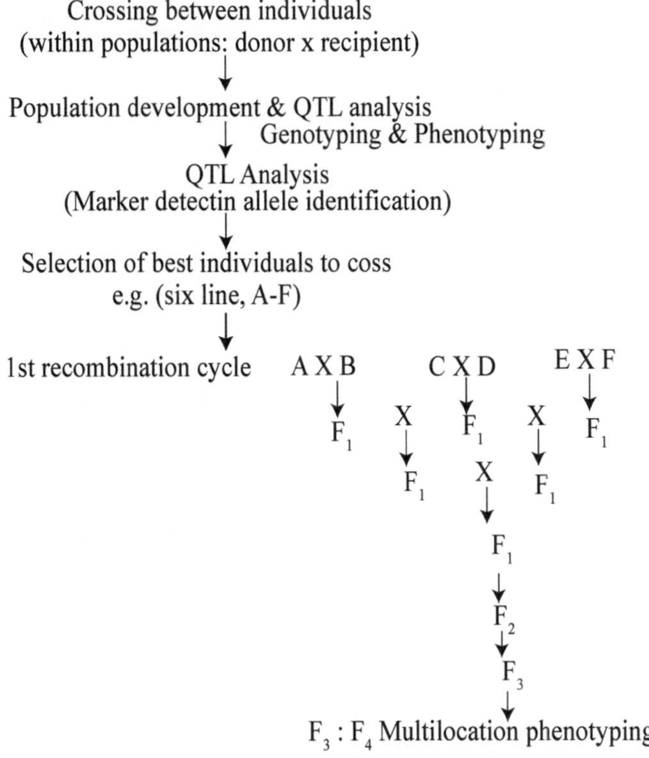

Fig 11: Steps in marker-assisted recurrent selection
Source: Dormatey *et al.*, 2020

Double Haploid Breeding

DHs are plants derived from a single immature pollen grain by androgenesis. It results in production of homozygous diploids after chromosome doubling of F1 individuals. By using DH populations, the genes of interest are immediately fixed through haplo-diploidization and homozygous gene combinations are generated from segregating material in only one generation. Selection for single gene is done in DH0 generation. Highly heritable traits are selected in DH1 generation. Selection for low heritable traits is done in DH2 & DH3 generation. The 'immortal' nature of DHs facilitates rapid and efficient identification of the genes for resistance.

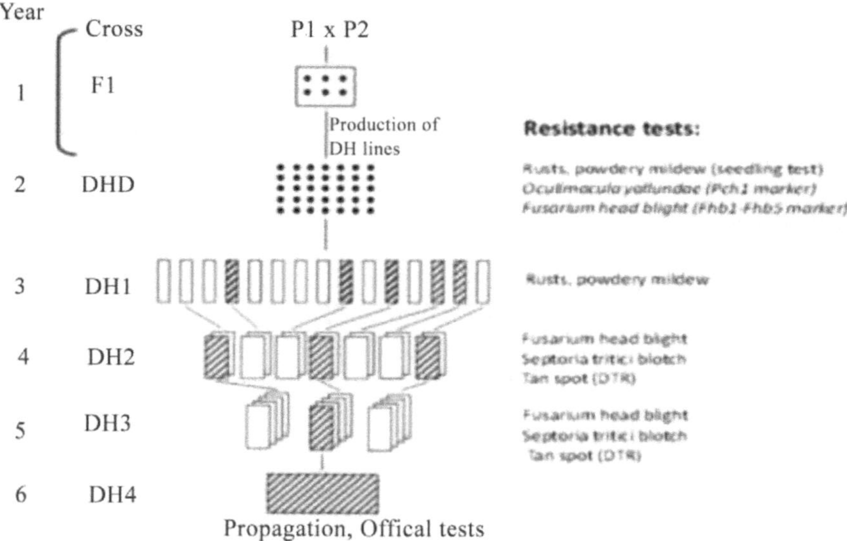

Fig. 11: Breeding scheme for self-pollinating crops using doubled haploid lines (DH)
Source- Miedaner, 2016

Genomic Selection

The basic idea behind GS approach is to use genome-wide marker data to effectively select elite individuals with superior traits. With the advent of SNP markers, GS is widely used. It uses phenotype and genotype data of past trials of lines/test populations or breeding population to predict the breeding values of individuals that have not been phenotyped, thus reducing the time required for screening superior individuals for breeding. The population used for both phenotype and genotype are called **training population (TRP)**, while the population having only genotype is called **breeding population (BP)** or test population (TSP). The BP is used to identify loci associated with the trait in

TRP. The idea is to search for associations among phenotype and genotype in a large, representative training population that was phenotyped in many environments and genotyped by whole-genome marker assays or sequencing technologies. A suitable statistical model showing association of genetic loci with phenotypes is created to estimate genomic estimated breeding values (GEBVs). The breeding value of breeding population is predicted based on its genetic relationship determined from the statistical model developed with the training population. Individuals from breeding population can be selected based on their breeding values for breeding purpose to speed up genetic gain. Facilitates recurrent selection and improve genetic gain from selection per unit time and cost, especially for quantitative disease resistances with low heritability. Thus, population size, nature of TRP, and relatedness of TRP to the BP are the main factors affecting GS efficacy. In this approach, the training population was selected from a random base population. This training population was phenotyped and genotyped in the first phase & model was predicted in the second phase. The best individuals are selected based on the highest breeding value for the desired trait. These breeding population are genotyped, and the previously predicted model is used for the evaluation of phenotype and top- ranking individuals are selected. Application - we have complete knowledge of phenotype and genotype that was tested and verified for field plantation and clonal propagation. One key advantage of using GS over MAS is that it considers all markers irrespective of their role in phenotype. The extent of genetic gains achieved per year is severalfold more using genomic selection than traditional breeding or MAS approaches. GS is advantageous over conventional breeding as it circumvents phenotypic evaluation and QTL mapping and thus reduces the time for enhancing genetic gains and overall cost (up to 50%) for selecting individuals for breeding and varietal development.

Speed Breeding

Speed breeding was first inspired by NASA and started by Dr. Lee Hickey. For most crop plants, the breeding of new, advanced cultivars takes several years which includes crossing of selected parent lines and 4–6 generations of inbreeding are generally required to develop genetically stable lines for evaluation of agronomic traits and yield. This is particularly time-consuming for field-grown crops that are often limited to only 1–2 generations per year. Hence, the concept of 'speed breeding' developed that involves extending photoperiod and controlled growing conditions such as temperature, soil media, spacing etc. in glasshouses to accelerate the developmental rate of plants enabling rapid generation advancement by shortening the breeding cycle & accelerating breeding and research programs. In RGA, as the name implies, the method can enable several generations or cycles (between F2 and F6 generations) within a short period compared to normal field conditions. The RGA method is similar to that of (SSD) method but in RGA growth conditions are manipulated to enforce early flowering and seed set compared to normal field conditions. This method rapidly generates fixed populations through SSD, which can be cheaper than generating double haploids & introgress genes or haplotypes into elite lines using marker-assisted selection.

RNAi Technology in Biotic Stress Resistance

To reduce losses caused by plant pathogens, plant biologists have devised numerous methods to engineer resistant plants. Among them, RNA silencing-based resistance has provided a powerful tool for engineering pest and disease-resistant plants in the last two decades (Werner *et al.*, 2020), opening new avenues for agrochemical design. The application of RNAi technology is based on the delivery of double-stranded RNA (dsRNA) or small interfering RNA (siRNA) for gene silencing. RNA interference (RNAi) in animals, is a conserved regulatory eukaryotic mechanism of gene expression that plays a crucial role in growth, development, and host defense (Maillard *et al.*, 2013). Thus, RNAi can be considered as a natural gene-based technology for highly specific pest control. The use of RNAi in pest management has been widely studied in different organisms, showing the utility of this technology in both basic and applied science. For instance, the expression of dsRNA-targeting genes coding for MAP Kinase and cyclophilin in wheat plants caused a severe reduction of leaf rust infection by *P. triticina* (Panwar *et al.*, 2018). Hence, the current application of dsRNA for pest and disease control is emerging as an appealing alternative to genetically modified crops (Marcianò *et al.*, 2021).

CRISPR Technology in Biotic Stress Resistance

Since the development of CRISPR/Cas (clustered regularly interspaced short palindromic repeats-CRISPR-associated) systems for precise genome editing, such as CRISPR/Cas9 and CRISPR/Cas12, there is a lot of potential to target susceptible genes (s genes) to enhance economically advantageous features in crops (Mubarik *et al.*, 2021). Stable site-specific mutagenesis can be induced without any pleotropic effect. This is a technology using RNA guided site-specific nuclease which can induce DNA lesion at the target region without the involvement of any gene transfer technology. This method has been utilised in many crops for disease resistance and especially s-gene-mediated resistance. The very first demonstration of utilisation of a novel CRISPR genome editing tool of S-gene mutation involved SWEET11 and SWEET14 in rice itself (Jiang *et al.*, 2013).

Discussion

Pest and disease-related yield losses can be 100%, as with rice blast (Fahad *et al*., 2019). Plant diseases can result in large losses in production if they strike at any point in the plant's growth cycle, but they can inflict the most harm if they strike at certain vulnerable and essential periods, such the seedling stage in the production of rice (Fahad *et al*., 2019). Although, diseases brought on by bacterial, viral, and fungal pathogens are economically significant in the production of cereals but impact of these diseases varies with location. This is because the virulence and aggression of certain pests and diseases are determined by their natural enemies, the type of crop being grown, the climate in the area, and the availability of the wild types. The most practical and economical way to address the present and potential future food supply-demand gaps is through crop improvement for biotic stress tolerance or resistance (Costa and Farrant, 2019). In Africa, numerous crop pest and disease problems (Gressel *et al*., 2004) and yield losses from biotic stresses (Cairns *et al*., 2013) have been resolved by plant breeding. New crop varieties are rapidly becoming outdated as a result of the rapid evolution of biological stresses brought on by climate change and natural selection. To quickly generate novel varieties, especially in maize hybrid production, modern breeding procedures such as double haploid technology, mutation breeding, and genomic selection are being applied (Prasanna *et al*., 2013). Crop improvement programs that use marker-assisted techniques to increase resistance to various insect pests and diseases include marker-assisted backcross breeding selection (MABS) and rapid generation advanced techniques which in turn produce new resistant varieties.

The accuracy of marker-trait association and the assessment of genomic breeding values are critical to the effectiveness of genomic-assisted breeding for plant stress response. These factors primarily depend on the coverage and accuracy of genotyping and phenotyping. Additionally, bacterial and viral infections can be countered by RNA interference technology to reduce production losses in grain crops. Understanding and enhancing cereal development and production under abiotic and biotic stresses has been made possible through the application of functional genomic research and molecular breeding. It is anticipated that the clustered, regularly interspaced, short palindromic repeats (CRISPR)/CRISPR-associated protein (Cas) system will be crucial to precision plant breeding and the creation of non-transgenic cereals resistant to biotic stressors and the negative effects of climate change.

Conclusion

Due to one of the crucial effects of climate change, several of the biotic stresses which were minor in the past have become major and affecting major cereal crops significantly. Breeding for climate- resilient cereal varieties with enhanced resistance to prevalent diseases and insect pests in changing climatic conditions involves identifying and incorporating genes associated with disease resistance from diverse genetic sources into breeding programs. Developing cultivars with natural resistance mechanisms against common pests like stem borers, leafhoppers, and aphids in rice is crucial by incorporating pest-resistant traits from wild relatives or through genetic engineering can enhance pest tolerance. Shortening the crop duration by breeding for early maturity can help mitigate the impacts of changing disease and pest dynamics. Early-maturing varieties can escape disease and pest pressures occurring later in the growing season, thereby reducing yield losses. Integrating multiple resilience traits into breeding programs through multi-trait selection approaches can enhance the overall resilience of cereal varieties to changing climatic conditions, diseases, and pests. Leveraging genomic technologies such as marker-assisted selection, genomic selection, and gene editing accelerates the breeding of climate-resilient cereal varieties by enabling precise trait introgression and selection.

References

Andargie, M., Li, L., Feng, A., Zhu, X., & Li, J. (2018). Mapping of the quantitative trait locus (QTL) conferring resistance to rice false smut disease. Current Plant Biology, 15, 38-43.

Bhardwaj, S. C., Singh, G. P., Gangwar, O. P., Prasad, P., & Kumar, S. (2019). Status of wheat rust research and progress in rust management-Indian context. Agronomy, 9(12), 892.

Boina, D., Prabhakar, S., Smith, C. M., Starkey, S., Zhu, L., Boyko, E., & Reese, J. C. (2005).
Categories of resistance to biotype I greenbugs (Homoptera: Aphididae) in wheat lines containing the greenbug resistance genes Gbx and Gby. Journal of the Kansas Entomological Society, 78(3), 252-260.

Boyko, E. V., S. R. Starkey, and C. M. Smith. 2004. Molecular genetic mapping of Gby, a new greenbug resistance gene in bread wheat. Theoretical and Applied Genetics 109:1230–1236.

Cairns, J. E., Crossa, J., Zaidi, P. H., Grudloyma, P., Sanchez, C., Araus, J. L., ... & Atlin, G. N. (2013). Identification of drought, heat, and combined drought and heat tolerant donors in maize. Crop Science, 53(4), 1335-1346.

Chander, S., Aggarwal, P. K., Kalra, N., & Swaruparani, D. N. (2003). Changes in pest profiles in rice- wheat cropping system in Indo-gangetic plains. Annals of Plant Protection Sciences, 11(2), 258- 263.

Chen G, Xiao Y, Dai S, Dai Z, Wang X, Li B, Jaqueth JS, Li W, Lai Z, Ding J, Yan J. Genetic basis of resistance to southern corn leaf blight in the maize multi-parent population and diversity panel. Plant Biotechnol J. 2023 Mar;21(3):506-520.

Chen, S., Wang, C., Yang, J., Chen, B., Wang, W., Su, J., ... & Zhu, X. (2020). Identification of the novel bacterial blight resistance gene Xa46 (t) by mapping and expression analysis of the rice mutant H120. Scientific reports, 10(1), 12642.

Collins, N., Drake, J., Ayliffe, M., Sun, Q., Ellis, J., Hulbert, S., & Pryor, T. (1999). Molecular characterization of the maize Rp1-D rust resistance haplotype and its mutants. The Plant Cell, 11(7), 1365-1376.

Costa, M. C. D., & Farrant, J. M. (2019). Plant resistance to abiotic stresses. Plants, 8(12), 553.

Devi, R. S., Reddy, A. V., & Dhurua, S. (2021). Reaction of gene differential rice varieties against gall midge, Orseolia oryzae (Wood-Mason) biotype at Warangal, India.

Dormatey, R., Sun, C., Ali, K., Coulter, J. A., Bi, Z., and Bai, J. (2020). Gene pyramiding for sustainable Crop improvement against biotic and abiotic stresses. Agronomy, 10(9): 1255.

Fahad, S., Rehman, A., Shahzad, B., Tanveer, M., Saud, S., Kamran, M., ... & ur Rahman, M. H. (2019). Rice responses and tolerance to m*et al*/m*et al*loid toxicity. In Advances in rice research for abiotic stress tolerance (pp. 299-312). Woodhead publishing.

Fiyaz, R. A., Yadav, A. K., Krishnan, S. G., Ellur, R. K., Bashyal, B. M., Grover, N., & Singh, A. K. (2016). Mapping quantitative trait loci responsible for resistance to Bakanae disease in rice. Rice, 9, 1-10.

Frey, L. A. (2023). Harnessing genetic diversity for improving southern anthracnose resistance and quality traits in red clover (Doctoral dissertation, ETH Zurich).

Gavhane, D. B., Kulwal, P. L., Kumbhar, S. D., Jadhav, A. S., & Sarawate, C. D. (2019). Cataloguing of blast resistance genes in landraces and breeding lines of rice from India. Journal of Genetics, 98(5), 106.

Gressel, J., Hanafi, A., Head, G., Marasas, W., Obilana, A. B., Ochanda, J., & Tzotzos, G. (2004). Major heretofore intractable biotic constraints to African food security that may be amenable to novel biotechnological solutions. Crop protection, 23(8), 661-689.

Hittalmani, S., Kumar, K. G., Shashidhar, H. E., & Vaishali, M. G. (2002). Identification of quantitative trait loci associated with sheath rot resistance (Sarocladium oryzae) and panicle exsertion in rice (Oryza sativa L.). Current Science, 82(2), 133-135.

Hore, T. K., Inabangan-Asilo, M. A., Wulandari, R., Latif, M. A., Nihad, S. A. I., Hernandez, J. E., ... & Swamy, B. M. (2022). Introgression of tsv1 improves tungro disease resistance of a rice variety BRRI dhan71. Scientific Reports, 12(1), 18820.

https://www.iari.res.in/files/Publication/Nilgiri_Wheat_News/NWN_11_23082023.pdf; retrieved on 22.04.2024

https://www.iari.res.in/files/Publication/Nilgiri_Wheat_News/NWN_06082018.pdf; retrieved on 22.04.2024

Hurni, S., Scheuermann, D., Krattinger, S. G., Kessel, B., Wicker, T., Herren, G., ... & Keller, B. (2015). The maize disease resistance gene Htn1 against northern corn leaf blight encodes a wall- associated receptor-like kinase. Proceedings of the National Academy of Sciences, 112(28), 8780-8785.

Jamann, T. M., Luo, X., Morales, L., Kolkman, J. M., Chung, C. L., & Nelson, R. J. (2016). A remorin gene is implicated in quantitative disease resistance in maize. Theoretical and Applied Genetics, 129, 591-602.

Jambagi, S. R., Kambrekar, D. N., Mallapur, C. P., & Naik, V. R. (2021). Status of shoot fly (Atherigona spp.) incidence in wheat in major wheat growing districts of North Karnataka. J Pharm Innov SP-10 (11), 2032-2035.

Jena, K.K., Kim, SM. Current Status of Brown Planthopper (BPH) Resistance and Genetics. Rice 3, 161–171 (2010).

Ji, Z., Wang, C., & Zhao, K. (2018). Rice routes of countering Xanthomonas oryzae. International journal of molecular sciences, 19(10), 3008.\

Jiang W, ZhouH, BiH, Fromm M, Yang B, Weeks DP. 2013. Demonstration of CRISPR/Cas9/ sgRNAmediated targeted gene modification in Arabidopsis, tobacco, sorghum and rice. Nucleic Acids Res. 41:e188

Kassa, M. T., Menzies, J. G., & McCartney, C. A. (2014). Mapping of the loose smut resistance gene Ut6 in wheat (Triticum aestivum L.). Molecular breeding, 33, 569-576.

Kosina, P., Reynolds, M., Dixon, J., & Joshi, A. (2007). Stakeholder perception of wheat production constraints, capacity building needs, and research partnerships in developing countries. Euphytica, 157, 475-483.

Kumar, Harish & Jakhar, Anil. (2018). Recent trends in insect-pests management of cereals crops.

Lee, S. B., Kim, N., Jo, S., Hur, Y. J., Lee, J. Y., Cho, J. H., & Park, D. S. (2021). Mapping of a major QTL, qBK1Z, for bakanae disease resistance in rice. Plants, 10(3), 434.

Leslie, J. F., Zeller, K. A., Logrieco, A., Mule, G., Moretti, A., & Ritieni, A. (2004). Species diversity of and toxin production by Gibberella fujikuroi species complex strains isolated from native prairie grasses in Kansas. Applied and Environmental Microbiology, 70(4), 2254-2262.

Li, D., Zhang, F., Pinson, S. R., Edwards, J. D., Jackson, A. K., Xia, X., & Eizenga, G. C. (2022). Assessment of rice sheath blight resistance including associations with plant architecture, as revealed by genome-wide association studies. Rice, 15(1), 31.

Li, Y., Li, Z., Cui, S., Chang, S. X., Jia, C., & Zhang, Q. (2019). A global synthesis of the effect of water and nitrogen input on maize (Zea mays) yield, water productivity and nitrogen use efficiency. Agricultural and Forest Meteorology, 268, 136-145.

Liu, Q., Liu, H., Gong, Y., Tao, Y., Jiang, L., Zuo, W., ... & Xu, M. (2017). An atypical thioredoxin imparts early resistance to sugarcane mosaic virus in maize. Molecular Plant, 10(3), 483-497.

Maas III, F. B., Patterson, F. L., Foster, J. E., & Hatchett, J. H. (1987). Expression and inheritance of resistance of 'Marquillo'wheat to Hessian fly biotype D. Crop Sci, 27(1), 49-52.

Mahapatro, G. K., & Kumar, S. (2013). First report of a commonly prevalent termite Heterotermes indicola from Delhi region. The Indian Journal of Agricultural Sciences, 83(4).

Maillard, P. V., Ciaudo, C., Marchais, A., Li, Y., Jay, F., Ding, S. W., & Voinnet, O. (2013). Antiviral RNA interference in mammalian cells. Science, 342(6155), 235-238.

Marcianò, D., Ricciardi, V., Marone Fassolo, E., Passera, A., Bianco, P. A., Failla, O., ... & Toffolatti, S. L. (2021). RNAi of a putative grapevine susceptibility gene as a possible downy mildew control strategy. Frontiers in Plant Science, 12, 667319.

Matsumoto, K., Ota, Y., Yamakawa, T., Ohno, T., Seta, S., Honda, Y., ... & Sato, H. (2021). Breeding and characterization of the world's first practical rice variety with resistance to brown spot (Bipolaris oryzae) bred using marker-assisted selection. Breeding science, 71(4), 474-483.

Menzies, J. G., Nielsen, J., Thomas, P. L., & Knox, R. E. (2003). Virulence of Canadian isolates of Ustilago tritici: 1964–1998, and the use of the geometric rule in understanding host differential complexity. Canadian journal of plant pathology, 25(1), 62-72.

Miedaner, T. (2016). Breeding strategies for improving plant resistance to diseases. In Advances in plant breeding strategies: Agronomic, abiotic and biotic stress traits. Springer, Cham., pp. 561- 599.

Molla, K. A., Karmakar, S., Molla, J., Bajaj, P., Varshney, R. K., Datta, S. K., *et al.* (2020). Understanding sheath blight resistance in rice: the road behind and the road ahead. Plant Biotechnol. J. 18 (4), 895–915.

Mubarik, M. S., Khan, S. H., and Sajjad, M. (2021). "Key Applications of CRISPR/Cas for Yield and Nutritional Improvement," in CRISPR Crops. Editors A. Ahmad, S. H. Khan, and Z. Khan (Singapore: Springer). doi:10.1007/978-981-15-7142-8_7

Padmavathi, G., Ram, T., Ramesh, K., Rao, Y. K., Pasalu, I. C., & Viraktamath, B. C. (2007). Genetics of whitebacked planthopper, Sogatella furcifera (Horváth), resistance in rice.

Panwar, V., Jordan, M., McCallum, B., & Bakkeren, G. (2018). Host-induced silencing of essential genes in Puccinia triticina through transgenic expression of RNA i sequences reduces severity of leaf rust infection in wheat. Plant biotechnology journal, 16(5), 1013-1023.

Paudel, B., Zhuang, Y., Galla, A., Dahal, S., Qiu, Y., Ma, A., ... & Yen, Y. (2020). WFhb1-1 plays an important role in resistance against Fusarium head blight in wheat. Scientific reports, 10(1), 7794.

Phi, C. N., Fujita, D., Yamagata, Y., Yoshimura, A., & Yasui, H. (2019). High-resolution mapping of GRH6, a gene from Oryza nivara (Sharma et Shastry) conferring resistance to green rice leafhopper (Nephotettix cincticeps Uhler). Breeding Science, 69(3), 439-446.

Prakash, A., Bentur, J. S., Prasad, M. S., Tanwar, R. K., Sharma, O. P., Bhagat, S., ... & Jeyakumar, P. (2014). Integrated pest management for rice. National Centre for Integrated Pest Management, New Delhi, India, 43.

Prasanna, B. M., Cairns, J., & Xu, Y. (2013). Genomic tools and strategies for breeding climate resilient cereals. In Genomics and Breeding for Climate-Resilient Crops: Vol. 1 Concepts and Strategies (pp. 213-239). Berlin, Heidelberg: Springer Berlin Heidelberg.

Rao, Y., Dong, G., Zeng, D., Hu, J., Zeng, L., Gao, Z., ... & Qian, Q. (2010). Genetic analysis of leaffolder resistance in rice. Journal of Genetics and Genomics, 37(5), 325-331.\

Ratcliffe, R. H. (2013). Breeding for Hessian fly resistance in wheat. Radcliffe's IPM World Textbook, University of Minnesota, St. Paul, MN, USA.

Satyagopal K, Sushil SN, Jeyakumar P, Shankar G, Sharma OP, Sain SK, Boina DR, Chattopadhyay D, Rao NS, Sunanda BS, Asre R, Murali R, Kapoor KS, Arya S, Kumar S, Patni CS, Sharma I, Ravindra H, Shivanna BK, Saharan MS, Kaur D, Singh B, Sharma RK, Chokar RS, Karnatak AK, Tiwari R, Deepshikha, Patel BR, Khalko S, Laskar N, Roy A, Hath TK, Ghetiya LV, Sathyanarayana N, Latha S. 2014. AESA based IPM package for Wheat. pp 58.

Sharma, A. K., Babu, K. S., Nagarajan, S., Singh, S. P., & Manoj, K. (2004). Distribution and status of termite damage to wheat crop in India. Indian Journal of Entomology, 66(3), 235-237.

Singh, P. K., Gahtyari, N. C., Roy, C., Roy, K. K., He, X., Tembo, B., ... & Chawade, A. (2021). Wheat blast: a disease spreading by intercontinental jumps and its management strategies. Frontiers in Plant Science, 12, 710707.

Stuart, Jeffrey J.; Chen, Ming-Shun; and Harris, Marion O., "Hessian Fly" (2008). Publications from USDA-ARS / UNL Faculty. 397. ht t ps:/ / di gi t al commo ns.unl.edu/ usdaa rsfac pub/ 397

Sun, H., Zhai, L., Teng, F., Li, Z., & Zhang, Z. (2021). qRgls1. 06, a major QTL conferring resistance to gray leaf spot disease in maize. The Crop Journal, 9(2), 342-350.

Tadesse, W., Bishaw, Z., & Assefa, S. (2018). Wheat production and breeding in Sub-Saharan Africa: Challenges and opportunities in the face of climate change. International Journal of Climate Change Strategies and Management, 11(5), 696-715.

Tadesse, W., El-Hanafi, S., El-Fakhouri, K., Imseg, I., Rachdad, F. E., El-Gataa, Z., & El Bouhssini, M. (2022). Wheat breeding for Hessian fly resistance at ICARDA. The Crop Journal, 10(6), 1528-1535.

Tekete, C., Cunnac, S., Doucouré, H., Dembele, M., Keita, I., Sarra, S., ... & Verdier, V. (2020). Characterization of new races of Xanthomonas oryzae pv. oryzae in Mali informs resistance gene deployment. Phytopathology, 110(2), 267-277.

Thambugala, D., Menzies, J. G., Knox, R. E., Campbell, H. L., & McCartney, C. A. (2020). Genetic analysis of loose smut (Ustilago tritici) resistance in Sonop spring wheat. BMC Plant Biology, 20(1), 314.

Weng, Y., and M. D. Lazar. 2002. Amplified fragment length polymorphism- and simple sequence repeat-based molecular tagging and mapping of greenbug resistance gene Gb3 in wheat. Plant Breeding 121:218–223.

Werner, B. T., Gaffar, F. Y., Schuemann, J., Biedenkopf, D., & Koch, A. M. (2020). RNA-spray-mediated silencing of Fusarium graminearum AGO and DCL genes improve barley disease resistance. Frontiers in Plant Science, 11, 476.

Wisser, R. J., Kolkman, J. M., Patzoldt, M. E., Holland, J. B., Yu, J., Krakowsky, M., ... & Balint- Kurti, P. J. (2011). Multivariate analysis of maize disease resistances suggests a pleiotropic genetic basis and implicates a GST gene. Proceedings of the National Academy of Sciences, 108(18), 7339-7344.

Xie, X., Chen, Z., Cao, J., Guan, H., Lin, D., Li, C., & Wu, W. (2014). Toward the positional cloning of qBlsr5a, a QTL underlying resistance to bacterial leaf streak, using overlapping sub- CSSLs in rice. PLoS One, 9(4), e95751.

Yadav, M. K., Aravindan, S., Ngangkham, U., Raghu, S., Prabhukarthikeyan, S. R., Keerthana, U., ... & Rath, P. C. (2019). Blast resistance in Indian rice landraces: Genetic dissection by gene specific markers. Plos one, 14(1), e0211061.

Zaidi, P. H., Rafique, S., & Singh, N. N. (2003). Response of maize (Zea mays L.) genotypes to excess soil moisture stress: morpho-physiological effects and basis of tolerance. European Journal of Agronomy, 19(3), 383-399.

Zheng, A., Lin, R., Zhang, D., Qin, P., Xu, L., Ai, P., *et al.* (2013). The evolution and pathogenic mechanisms of the rice sheath blight pathogen. Nat. Commun. 4, 1424.

Zhu, L., C. M. Smith, A. Fritz, E. V. Boyko, and M. B. Flinn. 2004. Genetic analysis and molecular mapping of a wheat gene conferring tolerance to the greenbug (Schizaphis graminum Rondani). Theoretical and Applied Genetics 109:289–293

Zuo, W., Chao, Q., Zhang, N., Ye, J., Tan, G., Li, B., & Xu, M. (2015). A maize wall-associated kinase confers quantitative resistance to head smut. Nature genetics, 47(2), 151-157.

5

Use of GIS Based Technology for Plant Disease and Pest Management

Satish Kumar Singh[1], Anurag Patel[2] and Himanshu Tripathi[3]

[1]*Senior Research Fellow, NICRA Project, ICAR- Central Institute of Agricultural Engineering, Bhopal-462010, Madhya Pradesh*
[2]*Assistant Professor, School of Agriculture, Sanjeev Agrawal Global Educational University, Bhopal-462022, Madhya Pradesh*
[3]*Assistant Professor, Department of Agricultural Engineering RSM Collage Dhampur, Bijnor - 246761, Uttar Pradesh*

Abstract

While the fourth agricultural revolution has been sparked by technological advancements, humans still face many challenges in achieving agricultural resilience to meet the demands of the world's expanding population, including declining croplands, depleting water supplies, the negative effects of climate change, and more. On the one hand, food and dairy products are in increasing demand due to the growing population. On the other hand, an increasing number of variables pose a threat to food security and sustainable food production, including globalization, climate change, scarce land and water resources, unsustainable land management techniques, pests and diseases becoming resistant to chemical pesticides, and the emergence of pathogens that have adapted to new hosts. The integration of geographic information systems (GIS), remote sensing, and artificial intelligence (AI) in geospatial approaches offers a solid foundation for the sustainable management of agricultural resources with the goal of boosting agricultural output. These days, local, regional, and worldwide agricultural production uses these sophisticated instruments more and more. Growers have been more interested in implementing precision agriculture during the past ten years. Precision agriculture is a crop-management method that takes into account the temporal and geographical heterogeneity of soil and crops within a field. Another agricultural technique that is being promoted as a future development that will enable the world's population to be fed sustainably is precision agriculture. The broad use of geographic methods for agricultural resource management, including crop disease monitoring and pest infestation control, is the main topic of this chapter.

Introduction

Agriculture is the backbone of the economy, particularly in developing countries, and plays a pivotal role in economic stability and ensuring food security. As a matter of fact, the global population is increasing exponentially and is expected to reach 9.8 billion by 2050. Despite environmental restrictions, this sharp rise in the world's population puts pressure on agricultural resources to provide more food. (Ghosh 2012, FAO 2017). Obstacles to extending agricultural land, water scarcity, and climate change are some of the main challenges to raising crop productivity. Therefore, the few fundamental measures for sustainable management of agricultural resources to fulfill future food demands are increased species diversity, conservation of soil quality, farmland management, and wise use of water (Gomiero, 2016). Most crucially, in order to achieve the goals of global food security, improved agricultural crop productivity and a decrease in the use of input resources are dependent on timely intervention tactics combined with crop growth and health monitoring. Recent technological developments have demonstrated significant advancements in data analysis tools, communication networks, and sensor technology. In order to achieve the aims of global food security, such technological advancements have made it easier to use agricultural input variables sustainably, mitigate potential losses, and promote improved and sustained production. Global food security now depends on a wide range of established and emerging methods and technologies, including Geographic Information Systems (GIS), Artificial Intelligence (AI), Remote Sensing (RS), Global Navigation Satellite Systems (GNSS), Internet of Things (IoT), and Big Data Analytics (BDA). These technological solutions are providing data-driven knowledge and insight for targeted or site-specific crop management, attempting to assure increased yield when combined with other sources of proof (Delgado *et al.* 2019: Satish *et al.* 2023) A geographic information system (GIS) is a kind of database system that combines tools for information collection, archiving, and retrieval with tools for evaluating, transforming, and presenting spatiotemporal data for a specific purpose. (Gomiero 2013; Koroleva 2014; Delgado *et al.* 2019; Satish *et al.* 2022)

A geographic information system (GIS) is a kind of database system that combines tools for information collection, archiving, and retrieval with tools for evaluating, transforming, and presenting spatiotemporal data for a specific purpose (Mathenge *et al.* 2022).

GIS is widely used in resource management in the food, industrial, and agricultural sectors as well as in environmental studies and medical services. Global, regional, and local agriculture production have been using GIS tools more and more in recent years. GIS has been extensively used to enable agricultural and soil interventions, along with other methods like RS, GNSS, and data analytics. (Delgado *et al.* 2019)

Using satellites in orbit or the sky, remote sensing is a kind of geospatial technology that gathers data on an object or phenomenon without the need for direct physical contact. Images of the earth's surface are taken using remote sensing techniques in various Electro- Magnetic Spectrum (EMS) wavelength ranges. A thermal infrared wavelength zone represents estimates of energy emitted by the earth's surface, while a microwave region represents a measure of relative return from the earth's surface. Some of the photos exhibit reflected solar radiation in the visible and near infrared regions. In order to use remote sensing technology, one must first produce electromagnetic radiation from the sun. This radiation strikes various things on Earth and is reflected into the atmosphere, where it is detected by a satellite sensor located far away and converted into images of various wavelengths. Nonetheless, (Figure 1) provides a summary of the procedure for gathering data on a subject or region of interest through the measurement of radiant energy reflected or emitted by a surface or object.

The Complete Processing of Remote Sensing Involves Various Steps Including

(1) Emission of electromagnetic radiations;

(2) Energy transmission from source to earth's surface;

(3) Interactions of EMR with the earth's surface, that is, reflection and emission;

(4) Energy transmission from the surface to satellite sensors; (5) Sensor data output; and

(6) Transmission, procession, and analysis of satellite data.

This remotely sensed data consists of both spectral (tone, color, and spectral signature) and spatial (size, shape, and orientation) information. All forms of data are created, managed, analyzed, and mapped by GIS, as is well known. GIS links information to a map by fusing location data which tells us where things are with various kinds of descriptive information which tells us what those things look like. In practically every business, this serves as a foundation for mapping and analysis, helping users to comprehend relationships, trends, and geographic context. Better management and decision-making, as well as increased efficiency and communication, are among the advantages. While remote sensing and geographic information systems (RS and GIS) have been extensively used in many agricultural domains, including soil mapping, terrain analysis, crop stress mapping, crop yield mapping, soil nutrient assessment, and mapping of organic matter, it has been suggested that high- resolution remotely sensed data is valuable for precision agriculture because it can deal with spatiotemporal variations. Different combinations of geographical resolutions, spectral coverage, and frequency ranges are used in precision agriculture. Low-resolution satellite data has reportedly been used to map crop production and monitor crop growth; fine- resolution satellite data is unavoidably needed to map the extent of disease infestation. Books (Gebeyehu 2019).

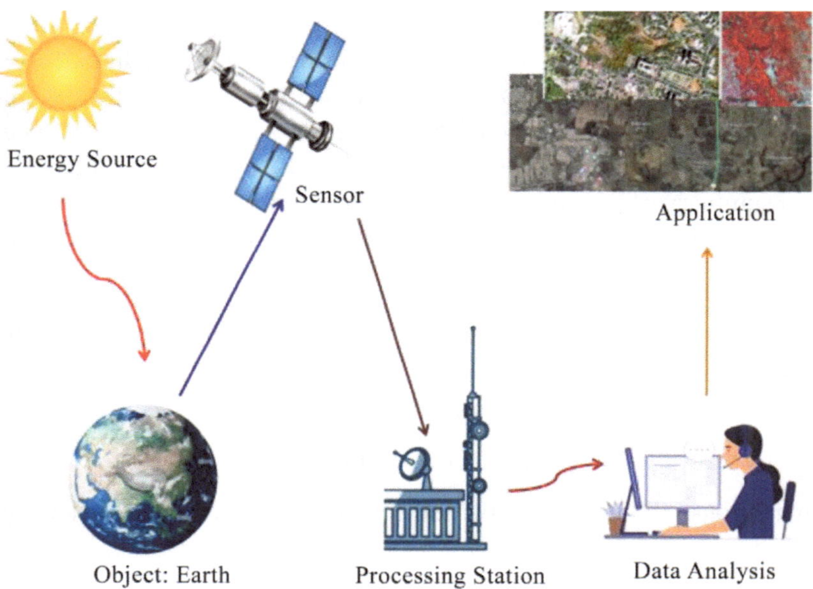

Fig. 1: Concept of remote sensing and GIS-applications of remote sensing. Electromagnetic energy

Electromagnetic energy travels in waves and covers a wide range of wavelengths, from long radio waves to short gamma rays (Figure 2). Only a small fraction of this spectrum, known as visible light, can be detected by the human eye. A radio detects a distinct part of the spectrum, whereas an x-ray equipment uses a different part of the spectrum. NASA's scientific instruments examine the Earth, the solar system, and the cosmos beyond using the entire electromagnetic spectrum.

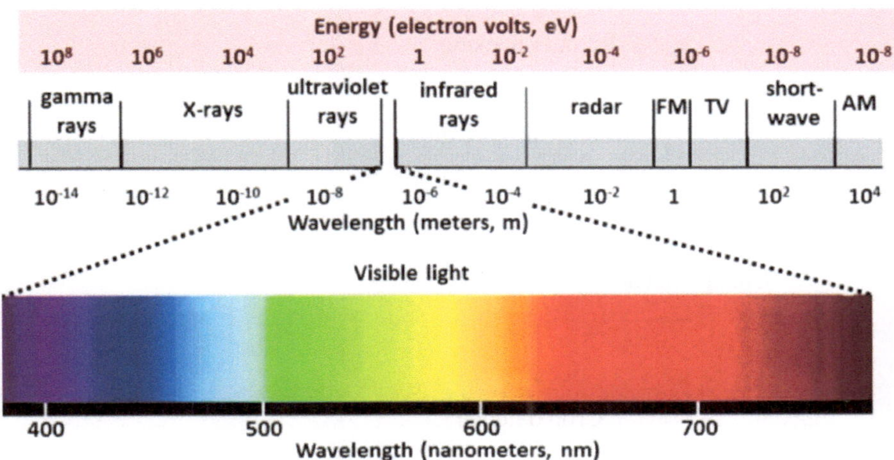

Fig. 2: Electromagnetic Spectrum

Source: https://www.google.com/search?sca_
esv=dbec97f3c192e658&q=electromagnetic+energy&uds

Pest Infestation and Crop Disease Management

Over the past few centuries, the mismatch between the increasing food demand of the world's booming population and the amount of agricultural supply has made the issue worse (Ramankutty *et al.* 2018). Both biotic and biotic restrictions are the primary limiting factors that might affect agricultural productivity, growth, and ultimately food security. A few of these are extreme and adverse weather, a scarcity of water, poor soil, insect infestation, diseases, and weeds. Furthermore, global agricultural productivity and output are adversely affected by temperature variations and other climatic disruptions, especially in tropical regions (Sare M. 2019). Experts estimate that by 2080, agricultural output will have decreased by 20% due to global climate change, with the situation in developing countries such as India expected to worsen much worse. Extreme weather events like floods and droughts are just a few instances of how agricultural production is being impacted by climate change. These changes and disturbances have led to an increase in the prevalence of disease outbreaks and insect pests. (Raza *et al.* 2019). As seen in Figure 3, GIS and remote sensing are being employed more than ever for accurate data gathering and analysis when it comes to crops. This section will examine the commendable role that geospatial technology has played in the identification and management of insect pests and plant/crop diseases. This section might possibly provide additional insight into how agricultural geospatial technology could be used to mitigate disease and insect-pest attacks.

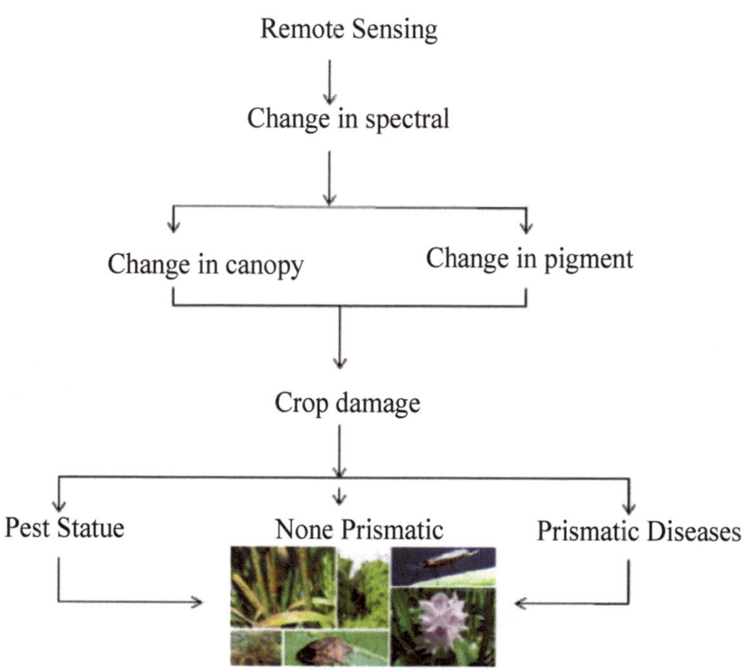

Fig. 3: Flowchart of crop-damaging diseases and pests, and the use of remote sensing.

A plant is considered "healthy" when its genetic potential for physiological impacts is being completely realized. The regular physiological processes of plants are disrupted when pests or diseases cause havoc (Sabtu *et al.* 2018). Insect-borne diseases and pests have a detrimental effect on the biomass, output, and overall health of the plant. Throughout the growing season, the effects of diseases and insect pests on quality and yield are typically immediately noticeable. Pests and illnesses that affect forests can have a significant impact on ecological services by reducing the forest's ability to sequester carbon, affecting biodiversity and wildlife habitat, changing natural landscaping and its cultural value, and negatively impacting agricultural productivity. An area of the earth's surface is described by data that is stored, captured, analyzed, and shown via GIS. Data storage and retrieval, presentation and querying, evaluation (geometric and thematic), conversion, and other characteristics are among the most used GIS functions. A GIS is constructed using four main pillars: information collection, processing, administration, and display (Anwer A, Singh G. 2019). The range of technical instruments at our disposal is commensurate with the growth of data- capture techniques. Plant health monitoring has become considerably easier as a result of developments in geospatial tools such as GPS, geodetic surveys, laser scanners, aerial photography, satellite-based remote sensing, and so on. GIS/GPS technology was utilized to evaluate semi-automated data in Arizona, Northern California, Idaho, Washington, and Oregon. (Zahra *et al.* 2023)

We employed an integrated approach to analyze satellite photos, plant growth data, and weather information in a GIS to determine the best growing conditions for our crops as well as the possibility of pests and illnesses. For agricultural locations, forecasts were created for favorable growth conditions as well as the development of illness using elevation, weather, and satellite data. Every day, models for twelve crop diseases and six insect pests were generated and presented on geo-referenced maps for agricultural survey areas. Remarkably accurate forecasts were also made for grape harvest dates and yields. Every day, information was shared online in the form of regional maps showing the risk of disease, insects, and weather, which were created using GIS data. (Anwer A, Singh G. 2019). Remote sensing has been applied to a wide range of problems, including weed management, crop forecasts, insect detection, water requirements, and plant nutrition challenges. Remote sensing can identify changes in plant properties as markers of disease and insect attack. The ability of a remote sensor to detect minute changes in vegetation is what sets it apart from traditional ground-based optical scouting approaches. This makes remote sensors useful for monitoring crop development, measuring within-field variability, and managing fields according to current conditions. (Zahra *et al.* 2023).

Geospatial technologies, such as GPS integration with GIS for Insect-Pest Management (IPM), are becoming essential to precision agriculture. The use of remote sensing, geographic analysis, and other technologies, such the Global Navigation Satellite System (GNSS), further enhances crop management planning (Ghosh and Kumpatla, 2022). These methods do have certain disadvantages, though. For instance, not all places, particularly rural ones, have easy access to high spatial resolution imaging. While hyper- spectral photography is critical, rapid mapping and nearly real-time picture collection and distribution systems are also critical to the success of this field of study. Consultancies and end-users/farmers frequently lack specialized expertise of geospatial techniques. Notwithstanding these challenges, geospatial technology can also yield valuable information in an integrated pest management (IPM) context. This is because it can help with a comprehensive understanding of the relative spatial features of a field and its crop, as well as provide data on the populations of pests and diseases that are either already present or anticipated (Anwer A, Singh G. 2019). This technique is proving beneficial in an area where large-scale pest and disease outbreak reports are increasing. The application of geographic technology has made it feasible to manage forestry more effectively in terms of production, disease and pest control, and other areas. Currently, academic settings are where these tools are most often employed, but this is expected to change as data and technology become more widely available. The broad use of geospatial data and technology has produced numerous benefits, two of which are higher yields and decreased worries about food insecurity (Zheng J., and Xu Y. 2023).

To ensure that everyone has access to enough food and water, regardless of where they reside, crop yield must increase in tandem with population expansion. In order to achieve agricultural sustainability, a range of technologies and approaches that increase production safety and lessen adverse environmental effects are used. Precision agriculture equipment are critical to agricultural success. Precision farming is a type of micromanagement that enhances agricultural and territorial management choices by utilizing data collected by geospatial technologies. Crop yields are heavily impacted by pests and diseases, which in turn impacts the world's capacity to feed its population. However, the advancement of geospatial technology has made it easier to control a wide range of pests and diseases that harm plants.

Conclusion

The utilization of geospatial data and technology is crucial in enhancing crop production and resolving concerns related to food security. To guarantee that everyone in a nation or region has access to enough nutrients, agricultural production must rise along with the population. In order to address the problem of pests and diseases that negatively impact plant health and significantly damage crop yield production which is necessary to feed the world's population this study has explored the function of geospatial technology. Geospatial technology has been applied to several activities, including organizing the gathered data and conducting initial crop health surveys. According to the review, a lot of research has used remote sensing technology to monitor the health of large groups of plants that are commodities for a nation, like oil palms, paddy, and wheat. However, less research has been done on crops that are planted on a small scale, like orchards. This could be because using the technology is expensive and may not be practical.

Consequently, in order to assist the ground team during the field survey and automatically record the position of impacted plants for future action, GPS technology has been deployed. In order to visualize the events from a large-scale perspective, the distribution of pest and disease incidence on the impacted crops has been mapped. The potential for an occurrence that is likely to occur in a few years has been predicted using the power of spatial analysis. The liberalization of international trade and climate change have posed significant obstacles for local crops in terms of yielding high-quality goods and ensuring food security for the global population. However, the development of geospatial technology has made it far simpler than ever before to combat the different pests and diseases that harm plant health.

References

Anwer A, and Singh G. 2019. Geo-spatial technology for plant disease and insect pest management. Bulletin of Environment, Pharmacology and Life Sciences.8:1-12

Delgado JA, Short NM, Roberts DP, and Vandenberg B. 2019. Big data analysis for sustainable agriculture on a geospatial cloud framework. Frontiers in Sustainable Food Systems.3:54-59.

Delgado JA, Vandenberg B, Neer DD, and Adamo R. 2019. Emerging nutrient management databases and networks of networks will have broad applicability in future machine learning and artificial intelligence applications in soil and water conservation. Journal of Soil and Water Conservation.74(6):113A-118A

FAO 2017. The Future of Food and Agriculture: Trends and Challenges. Rome: FAO Publishing. 4:1951-1960

Gebeyehu MN. 2019. Remote sensing and GIS application in agriculture and natural resource management. International Journal of Environmental Sciences & Natural Resources.19(2):45-49

Ghosh P, and Kumpatla SP. 2022. GIS applications in agriculture. In Geographic Information Systems and Applications in Coastal Studies. Intech Open.

Ghosh P, Kumpatla SP. 2012. GIS applications in agriculture. In: Zhang Y, Cheng Q, editors. Geographic Information Systems and Applications in Coastal Studies. London, UK: Intech Open; pp. 1-27

Gomiero T. 2013. Alternative land management strategies and their impact on soil conservation. Agriculture. 3(3):464-483

Gomiero T. 2016. Soil degradation, land scarcity and food security: Reviewing a complex challenge. Sustainability. 8(3):1-41

Koroleva EV, and Nikitin YY. 2014. U-max-statistics and limit theorems for perimeters and areas of random polygons. Journal of Multivariate Analysis. 127:98-111

Mathenge M, Sonneveld BGJS, and Broerse JEW. 2022. Application of GIS in agriculture in promoting evidence-informed decision making for improving agriculture: A systematic review. Sustainability.14(16):9974-9999

Ramankutty N, Mehrabi Z, Waha K, Jarvis L, Kremen C, Herrero M, and Rieseberg LH. 2018. Trends in global agricultural land use: implications for environmental health and food security. Annual review of plant biology, 69, 789-815.

Raza A, Razzaq A, Mehmood SS, Zou X, Zhang X, Lv Y, and Xu J. 2019. Impact of climate change on crops adaptation and strategies to tackle its outcome: A review. Plants, 8(2), 34.

Sabtu NM, Idris NH, and Ishak MHI. 2018. The role of geospatial in plant pests and diseases: An overview. In IOP Conference Series: Earth and Environmental Science.169(1):012013

Sare M. 2019. Climate change and agricultural production. Culture and Environment.4:201-208.

Satish Kumar Singh, Bhupendra Singh Parmar, Sachin Gajendra, Anurag Patel and Neha

Kushwaha, 2023. Spectral Response at Different Water Depth Standing in the Field. Int. J. Plant Soil Sci., vol. 35, no. 21, pp. 1184-1191.

Satish Kumar Singh, Neha Singh Kushwha and Anurag Patel (2022). Remote Sensing and GIS Technologies in use of Soil and Water Conservation, Book chapter, Advanced Innovative Technologies in Agricultural Engineering for Sustainable Agriculture, AkiNik Publications, 2022, pp. 89-114.

Walker, Robert J. 2016" Population growth and its implications for global security." American Journal of Economics and Sociology 75, no. 4: 980-1004.

Zahra SM, Shahid MA, MaqboolZ, Sabir RM, Safdar M, Majeed MD, and Sarwar A. 2023. Application of Geospatial Techniques in Agricultural Resource Management.

Zheng J, and Xu Y. 2023. A Review: Development of Plant Protection Methods and Advances in Pesticide Application Technology in Agro-Forestry Production. Agriculture, 13(11), 2165.

6

Nano-Tech Approaches in Plant Disease Management

Rajesh Kumari[1], Devesh Pathak[2] and Ravi Dhanker[3]

[1]Department of Plant Protection, Faculty of Agricultural Sciences, Aligarh Muslim University, Aligarh-202002, Uttar Pradesh, India
[2]Department of Plant Pathology, Faculty of Agricultural Sciences R.S.M. College, Dhampur, Bijnor-246761, Uttar Pradesh
[3]Department of Soil Science and Agricultural Chemistry, Faculty of Agricultural Sciences, R.S.M. College, Dhampur, Bijnor-246761 Uttar Pradesh

Abstract

Plant disease management technique applicable these days are predominantly relies on hazardous synthetic chemicals that are efficiently deleterious to the human and environment as well. Crop loss every year due to plant pathogen and pest is ranging 30–50%. Nanotechnology, proven advantageous and can offer positive environmental impacts over these harmful chemical pesticides by reducing toxicity, increasing the feasibility of pesticides which are normally baffling in water and improving longevity. Now a days and in coming year, it becoming one of the rapidly growing, advancing and most fascinating technological sciences in many disciplines like medicine, science, technology and more importantly in agriculture, to protect crop from different kinds of pest and pathogen on which most of India's population is completely dependent. This chapter briefly describe the introduction, role and impact of nanotech in management of plant disease. It is considered as modern technology in facing these widespread destructive challenges and causes in controlling in plant disease. In this chapter we tried to summarize the importance of nanotech in sustainable agriculture and application of various nanoparticles (NPs) in the management of plant disease. NPs can be produced by distinct practices, biological and chemical are one of them important methods. Biological method is widely and commercially used. Also, shortly describe the historical background for some NPs types. This chapter briefly focus on the synthesis of some compounds of NPs, for example silver NPs, silver and silica silver etc and their influence on plant disease also briefly describe the working principle of AgNPs against microorganism. NPs may act in accordance with chemical pesticides. Because of microscopic size, NPs have wide scope in virus control in plant. Now day's biosensors are used for the quick and precise pathogen detection and diagnosis. Regardless of immense scope of NPs application in managing the diseases of plant, there are precise apprehensions, weakness and risk associated with the utility of NPs in agriculture. It is necessary to be figured out on primacy basis prior to trade of NPs in agriculture.

Introduction

Nanotechnology, at the atomic and molecular level, is research and technology having 1-100 nm range. It is used to synthesize various forms of molecular and cellular structures in the body such as enzymes, proteins, carbohydrates, and lipids. In the Past decades, it is growing as interdisciplinary science that link with different material science, biology, chemistry, physics and engineering (Prasad 2004; Islam and Miyazaki 2009). In the world, It is emerged as fifth generation electronics that utilizes NPs combinations in instruments and equipment with the various dimensions of nanoscale.

Globally, nanotechnology attracts attention due to unique properties of nano size particles (Gajjar *et al.* 2009; Gong *et al.* 2007; Rai *et al.* 2009). Prasad *et al.* 2014, 2016, 2017 suggested that Nanotechnology have various applications, like improvement in energy and agriculture production and treatment of drinking water, the most important in disease diagnosis, processing of food and storage, environment pollution (mainly air pollution), monitoring of health, and pest and insect resistance. Agriculture is considered as economic sector as it provide food for increasing population, but due to environment variations this sector is facing many challenges (Anonymous 2009). Globally plant diseases are major constraint in the food production. In the list of Nanotechnology uses, Agriculture occupies second place. It helps to increase the soil fertility and thus the crop yields by decreasing the environmental threat and challenges through production of pesticides, synthetic fertilizers and managing the plant disease by the use of nanoparticles (Bhattacharyya *et al.* 2016; Singh *et al.* 2014).

For sustainable agriculture, control of plant disease is an important aspect in order to save limited food resources. There are many diseases which causes threat to food resource for future generation. Some devastating diseases are rice blast, late blight of potatoes, and brown spot of rice also called Bengal famine, chestnut blight, and citrus canker. However, control of plant disease is very difficult under various conditions. Many different methods and strategies have been used to achieve control of plant disease, like crop rotation, resistant cultivars, cultivation of pathogen free seeds and seedlings, proper spacing and time, controlled field moisture and pesticides use. Globally, plant diseases causes up to 10% yield reduction, it may exceeds 20%, whereas some diseases cause severe economic loss to the farmers. In order to solve the related problems, advanced and new technology is needed to combat the prevailing issues of plant diseases. Nanotechnology is effective modern technology and play important role in controlling the plant disease and proved to be the best alternative to the pesticides use especially with application of NPs (Martinelli *et al.* 2014) in managing the diseases of plant. Nanotechnology has valuable and effective practices. It involves the application of NPs on seeds, soil and leaves and provides protection to the plants from the pest and pathogens and thus control initiation and spread of infection (Khan and Rizvi 2014) especially against bacterial and fungal pathogens. NPs have high reactivity against microorganism due to their ultra-small size, smaller than a virus.

Synthesis of Nanoparticle

To present an overall nanotechnology picture below summarize different methods of NPs synthesis. The nanoparticles are synthesised mainly in two different ways bottom up and top down as shown below (Figure. 1). Chemical, physical and biological methods are another way of NPs synthesis. Its properties and efficiency of synthesis vary with the procedure. NPs synthesized more efficiently by chemical methods than other methods. Reduction, colloidal and sonochemical method are few examples of chemical methods.

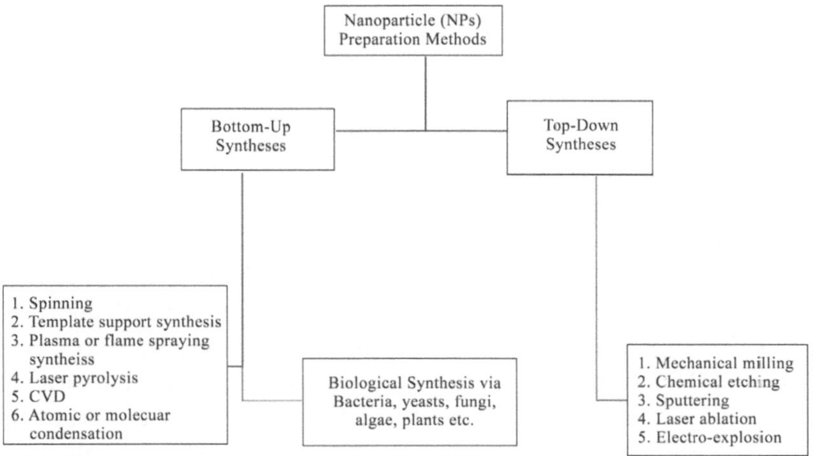

Fig. 1: Synthesis of Nanoparticles by two different ways (a) top-down and (b) bottom-up approaches.

Nanoparticle Type for Plant Disease Management

Rafique *et al.*, 2017 reported that silver NPs have gained popularity recently. Krishnaraj *et al.*, 2012 suggested that NPs have shown antifungal inhibition effect by well diffusion assay against various fungus such as of *Sclerotinia sclerotiorum, Botrytis cinerea, Macrophomina phaseolina, Alternaria alternata, Rhizoctonia solani* and *Curvularia lunata*. Another scientist Jain *et al* (2014) also showed sun-hemp rosette virus suppression when the nanoparticles of silver were sprayed on to leaves of bean crop. Further, the plants of faba bean infected with virus (yellow bean mosaic virus), and sprayed with silver NPs, produced significantly better results when the NPs were applied 24 hours post-infection. Silver NPs have shown immense potential in managing the plant disease against bacterial and fungal pathogens infection. Kah *et al.*, 2014, Mishra *et al.*, 2015. Foliar application of NP Cuo, Cu_2O, control late blight of disease caused by the fungal pathogen (*Phytophthora infestans*). Plant treated at 150-340 ug/ml CuO NPs performed more efficiently over other treatments. Bacterial spot caused by *Xanthomonas perforans* can be managed by copper and ammonium copper. The nano-copper products remarkably reduced the severity of bacterial spot disease in the green house and field. Higher amount of active Cu2+ were released from these composites which efficiently help in penetration of bacterial membrane. Now a days commercial Cu-based products and copper NPs products were equal, did not cause phytotoxicity and delivered less Cu per hectare. There are few essential plant micronutrient, copper is one of them which plays necessary role in growth and defence against plant diseases as well.

Nanoparticles that Act as Carriers

In order to develop effective agricultural formulations, NPs are commonly used as carrier to encapsulate, entrap, absorb and attach active molecules. Given below, there is brief discussion of NPs which have been used as carriers for insecticides, fungicides, herbicides and RNAi-inducing molecules. Mody *et al.* 2014 reported that silica NPs is considered as highly advantageous delivery vehicles and can be prepared or synthesized easily with a suitable and controlled size, shape and structure. They are pore-like holes that are produced in a spherical shape, for instance, porous hollow silica NPs (PHSNPs) or mesoporous silica NPs (MSNPs). Loading of pesticides inside the inner core by PHSNPs and MSNPs is suitable in order to provide a continuous release and to protect the active molecules. The PHSNPs shell structure provides protection to the active molecules inside NPs from degradation by ultra-violet light. Barik *et al.* 2008 showed that silica NPs seems to be the original (natural) choice for the development of product called agro-product against pest and pathogens also proved that plant tolerance can be enhance against various abiotic and biotic stresses by the use of silica NPs. Resistance in plant tissue against virus (mosaic virus) can be induced by chitosan ,which is a another popular NPs, by protecting them against infection caused by mosaic virus in various plant variety such as alfalfa, potato, snuff, cucumber and peanut [Kochkina *et al.*, 1994, Chirkov *et al.*, 2002]. NPs, chitosan have shown antimicrobial properties against Fusarium crown , tomato root rot, Botrytis bunch rot in grapes, *Pyricularia grisea* in rice [Kashyap *et al.*, 2015], but are less effective against bacteria [Malerba *et al.*, 2016].

Types of nanocarriers

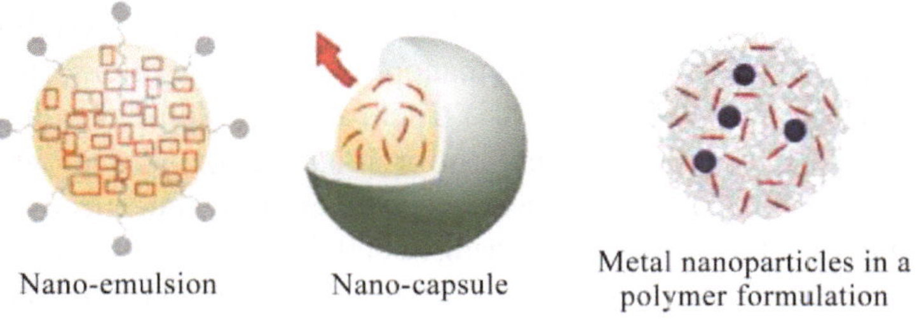

Fig. 2: Commonly used biodegradable polymers incorporated with nanopesticides (Source: Savita Rani and Sushil 2018).

The nanocarrier-active ingredient availability depends on organisms/ cells considered and the properties of carrier. Moreover, it have limited bioavailability, while its efficacy and toxicity depends on release of the active ingredient. However, as yet no information is available on nanopesticides availability. A polysaccharide, chitosan, commonly used as a polymer carrier for NPs. It has been reported by Wen *et al.,* (2010) that chitosan have the ability to change the enantioselective bioavailability of chiral herbicide dichlorprop. Protection and bioavailability of AI against photodegradation is completely dependent on its distribution within polymeric matrix suggested by Qing *et al.,* 2013

Nanopesticides (NPs)

NPs is any formulation that consist of element of nm size. Some of NPs, for the several years have already been on the market. It is not a single category product rather encompass a great variety of products. NPs includes variety of organic and inorganic ingredients in different forms. The main aim of nanoformulations are generally given below:

1. Improving the solubility of active ingredient which are normally poorly soluble.

2. Protection of active ingredient against premature degradation and releasing it in a targeted manner.

Xu *et al.*, 2010 suggested that better pesticide delivery systems can be achieved with the smaller size, optical transparency and low viscosity of nanoemulsion. Bioavailability and solubility of chemical active ingredient can be enhanced by micro and nano emulsion. According to Bergeson, 2010b NPs consist of small size of pesticide's active ingredients particles with useful pesticidal properties. It can enhance undesirable or unwanted pesticidal movement and agricultural formulations wettability and dispersion (Bergeson, 2010a). According to Bouwmeester *et al.*, 2009; Bordes *et al.*, 2009, there is need of NPs formulation in order to enhance the properties of nanomaterials such as solubility, thermal stability, permeability, biodegradability, crystallinity. Affinity to the target can be increased by nanopesticide formulations offering large specific surface area (Yan *et al.*, 2005). There are some delivery technique of NPs which can improve effectiveness in programmes of plant protection such as nanoemulsions, nanocontainers, nanocages and nanoencapsulates (Bouwmeester *et al.*, 2009; Lyons and Scrinis, 2009; Bergeson, 2010b). With a view to achieve slow release of NPK fertilizer, highly degradable antibacterial material i.e chitosan NPs have been proved beneficial. (Corradini *et al.*, 2010). Now a days, various form of formulation of NPs are available. Conventional pesticide formulations has hazardous effect., New type of formulations has been developed to reduce these effects with proper efficacy. One of the most capable is the use of nanotechnology based formulations i.e. nanoformulations.

Table 1: Efficacy of nano-formulations, Kah and Haffman (2014)

Type and Active Ingredient	**Comparison of efficacy of nanoformulation with that of commercial formulation**	**References**
Silver	Fungicidal activity against 18 plant pathogens (Petri dishes)	Mondal and Mani (2012)
Aluminium	Greater or similar insecticide activity than commercially available diatomaceous earth formulation	Stadler *et al*, (2010, 2012)
Chlorfenapyr Organic AI + Nano inorganic	Formulation of silica nanoparticle showed insecticidal activity twice as high than that of microparticles (laboratory and field tests)	Song *et al.*,(2012)
Silica Inorganic nano as active ingredient	For the stored grains, efficacy at similar rates than commercial diatomaceous earth (0.5-2 g/kg). Effect of size and coating to be further investigated	Debnath *et al.*, (2011, 2012)
TiO_2	Compared to standard treatment (greenhouse and field trials), better or on par efficacy	Paret *et al.*, (2013a,b)
Copper	Greater efficacy than Cu- oxychloride (in vitro suppression of bacterial blight of pomegranate at 0.2 mg/L and 2500-3000, respectively)	Kim *et al.*, (2012)

Mechanism of Nanoformulation Release

According to Kratz *et al.* (2012) "NPs start working after they are placed in a desired location". Chemical nature of formulation generally decides the release of bioactive compound. Controlled release of various polymeric nanomaterials proceeds through diffusion. According to Allan *et al.* (1971), a chemical interaction must be broken in order to release nanoformulation. Polymer insecticide bound in a chain reaction usually occurs via hydrolysis reaction. The control release depends on those chemical bound strength, whose properties of both the size and structure of macromolecules formed.

Management of Plant Diseases with Application of Different Types of Nanoparticle

With a view to achieve control on plant pest, pathogens and their vectors, recent advances in the production, design and fundamental understanding of nanomaterials are in trend. Compared to any other inorganic antibacterial agent properties, antimicrobial properties of silver has been investigated and applied thoroughly (Russell and Hugo 1994).

Plant Disease Control by Using Silver NPs

Antimicrobial effect of silver NPs are in both form i.e ionic and NPs form. It has been applied against pest and pathogens in a compound form or pure free m*et al* and it is non toxic to the humans (Yeo *et al.,* 2003; Elchiguerra *et al.,* 2005). Compared with synthetic fungicides, it is more safer in controlling several plant pathogens (Park *et al.* 2006). Multiple inhibitory modes of action against microorganisms is displayed by it (Clement and Jarret 1994) thus effectively manages the diseases. It shows inhibition on the colony formation of *Magnaporthe grisea* and *Bipolaris sorokiniana.*

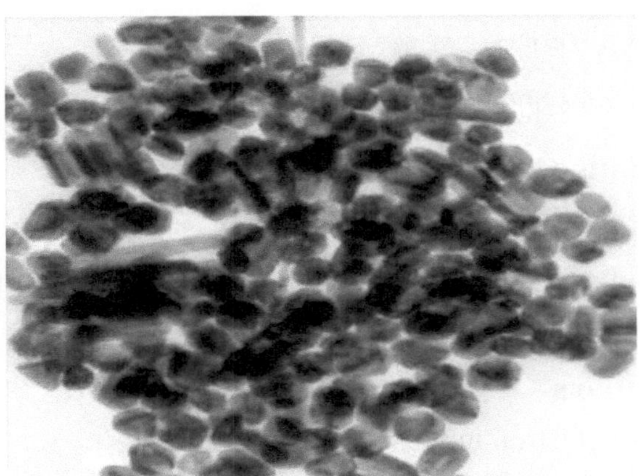

Fig. 3: Transmission of electron micrograph of collection of silver NPs, 200nm (http://pubs.rsc.org/en/content/articlehtml/2010/md/c0md0069h).

Silver and Silica-Silver

Particle's properties such as size, shape, types of compounds, exposure time and target too are some important basis on which efficacy of silver NPs is dependent. Fungal growth effectively suppressed by small size of nano silver. Particles of 1-5 nm sized may pass via protoplasmic membrane and fungi absorbed silica well into it (Wainwright *et al.,* 1986).

Disinfectant activity of NPs (silver) will increase as soon as absorption of silica-silver NPs occur into the fungal cells. Silica form physical barrier to pathogenic fungi thus increasing resistance to the disease. Control various diseases achieved by the application of silica-silver at lower than 2.5 ppm. The pathogen suppression usually not accomplish at this concentration on agar medium. Controlled of fungi by silica-silver occur and the fungi includes: *Sphaerotheca* spp., *Blumeria* spp., *Rhizoctonia* spp., *Phytopthora* spp., *Colletotrichum* spp., *Magnaporthe* spp., *Botrytis* spp. and *Pythium* spp.

Mode of Action of AgNPs Against Microorganism

Cell wall and membrane interaction

(a) Permeability changes

(b) phosphate administration disturbance

(c) Plasma membrane degradation (d) Proton motive force collapse (e) ATP synthesis inhibition

Impact on DNA and RNA

(a) Breaks in hydrogen bonding

(b) Nitrogen bases synthesis inhibition

(c) DNA and RNA synthesis disorders

(d) Inhibition of protein synthesis and denaturation of ribosomes

Influence on enzymes and amino acids

(a) Amino acid bonding, especially to –SH group

(b) Enzyme activity inhibition occur with the bonding to the active site of enzyme

Energy recruitment obstructions

(a) Influence on movement of electron in the respiratory chain

(b) Cytochrome inhibition

(c) Generating reactive oxygen species (ROS)

Adverse Effect and Risk Associated with NPS Applications in Disease Of Plant

Side effect and risk may be associated with the application of NPs. With a view to assess risk in the food chain and environment it require risk assessment strategies. Due to its unique properties they are used for environmental remediation. In redox reaction, NPs of colloidal iron have potential to perform as catalyst. For understanding the behaviour of NPs, risk assessments are needed to assess the risk associated with NPs use for remediation.

NPs possess several hazardous effects to environment, human beings and plants which is our major concern. Now a days, NPs are available in various form such as nanofertilizers, nanopesticides and other nanoformulations. They may deposit on leaves and other floral parts of plant when become air borne and may plug stomata and formed layer of toxic barrier on stigma which create hindrance in germination of pollen and penetration of tube into stigma. They may impair translocation of photosynthates, water and minerals after entering into the vascular tissue of plant. The NPs, when inhaled by human beings deep into the lungs during application results into various disorders and ill effects.

Conclusion

Worldwide, Agriculture sector plays important role in increasing the economy of the country and considered as important aspect in order to feed increasing population. Unfortunately it is facing threat in food production due to several devastating plant diseases and aberrant weather condition. It is important to select a reproducible and reliable system to conduct efficient studies at the cell, organism and ecosystem levels of the host-pathogen with a view to achieve effective control in plant disease. Several methods are employed to protect food in order to maintain sustainable agriculture. Among them nanotechnology has the potential to revolutionize the existing technologies used in plant disease management and can provide best solutions for agricultural applications. Nanotechnology, emerge as a useful science and has a wide scope of application in the field of agriculture in managing the disease for the effective exploitation. Choice of effective or suitable NPs based on some properties such as functional groups, morphology, size, and active loading capacity.

NPs of Au, Ti, Ag, Co etc suitable for coating enzyme based biosensors. Precise and quick diagnosis of infection in plant and detection of pesticides residue can be effectively done by these enzyme based biosensor. Despite of tremendous scope of nanotechnology application in managing the disease in plant, there are certain risk, demerits and apprehensions associated with the use of NPs in agriculture. Before the commercial use of nanotechnology in agriculture, it require to be worked out on priority basis. Concrete platform can be achieved via collaborative and multidisciplinary approach to bring the nanotechnology application for the plant protection in reality.

References

Allan, G. G., Chopra, C. S., Neogi, A. N., & Wilkins, R. M. (1971). Design and synthesis of controlled release pesticide-polymer combinations. Nature, 234, 349-351.

Anonymous. (2009). Nanotechnology and nanoscience applications: revolution in India and beyond. Strateg Appl Integrating Nano Sci.

Barik, T., Sahu, B., & Swain, V. (2008). Nanosilica—from medicine to pest control. Journal of Parasitology Research, 103, 253–258.

Bergeson, L. L. (2010). Nanosilver pesticide products: What does the future hold? Enviroment Quality Management, 19, 73-82.

Bergeson, L. L. (2010). Nanosilver: US EPA's pesticide office considers how best to proceed. Enviroment Quality Management, 19, 79-85.

Bhattacharyya, A., Duraisamy, P., Govindarajan, M., Buhroo, A. A., & Prasad, R. (2016).

Nano biofungicides: emerging trend in insect pest control. In: Prasad R (ed) Advances and applications through fungal nanobiotechnology. Springer International Publishing, Switzerland, 307–319.

Bordes, P., Pollet, E., & Averous, L. (2009). Nano-biocomposites: Biodegradable polyester/ nanoclay systems. Progress in Polymer Science, 34,125-155.

Bouwmeester, H., Dekkers, S., Noordam, M. Y., Hagens, W. I., & Bulder, A. S. (2009). Review of health safety aspects of nanotechnologies in food production. Regulatory Toxicology and Pharmacology. 53, 52-62.

Chirkov, S. (2002). The antiviral activity of chitosan (review). Applied Biochemistry and Microbiology, 38, 1–8.

Clement, J. L., & Jarret, P. S. (1994). Antimicrobial silver. M*et al*-Based Drugs 1, 467– 482.

Corradini, E., De Moura, M. R., & Mattoso, L. H. C. (2010). A preliminary study of the incorparation of NPK fertilizer into chitosan nanoparticles. Express Polymer Letters, 4, 509-515.

Elchiguerra, J. L., Burt, J. L., Morones, J. R., Camacho Bragado, A., Gao, X., Lara, H. H., & Yacaman, M. J. (2005). Interaction of silver nanoparticles with HIV-1. Journal of Nanobiotechnology 3, 6.

Gajjar, P., Pettee, B., Britt, D. W., Huang, W., Johnson, W. P., & Anderson, A. J. (2009). Antimicrobial activities of commercial nanoparticles against an environmental soil microbe, Pseudomonas putida KT2440. Journal of Biological Engineering, 3, 9.

Gong, P., Li, H., He, X., Wang, K., Hu, J., Zhang, S., &Yang X. (2007). Preparation and antibacterial activity of Fe3O4@Ag nanoparticles. Nanotechnology. 18, 604–611.

Islam, N., & Miyazaki, K. (2009). Nanotechnnology innovation system: understanding hidden dynamics of nanoscience fusion trajectories. Technological Forecasting and Social Change, 76, 128140.

Jain, D., & Kothari, S. (2014). Green synthesis of silver nanoparticles and their application in plant virus inhibition. Journal of Mycology Plant Pathohology, 44, 21.

Kah, M., & Hofmann, T. (2014). Nanopesticide research: Current trends and future priorities. Environment International, 63, 224–235.

Kashyap, P. L., Xiang, X., & Heiden, P. (2015). Chitosan nanoparticle based delivery systems for sustainable agriculture. International Journal of Boilogical Macromolecules, 77, 36–51.

Khan, M. R., & Rizvi, T. F. (2014). Nanotechnology: scope and application in plant disease management. Plant Pathology Journal, 13(3), 214–231.

Kochkina, Z., Pospeshny, G., & Chirkov, S. (1994). Inhibition by chitosan of productive infection of T-series bacteriophages in the Escherichia coli culture. Microbiologia. 64, 211–215.

Kratz, K., Narasimhan, A., Tangirala, R., Moon, S. C., Revanur, R., Kundu, S., Kim, H. S., Crosby, A. J., Russell, T. P., Emrick, T., Kolmakov, G., & Balazs, A. C. (2012). Probing and Repairing Damaged Surfaces with Nanoparticle-Containing Microcapsules. Nature Nanotechnology, 7, 87-90.

Krishnaraj, C., Ramachandran, R., Mohan, K., & Kalaichelvan, P. (2012). Optimization for rapid synthesis of silver nanoparticles and its effect on phytopathogenic fungi. Spectrochim. Acta Part A Molecular and Biomolecular Spectroscopy, 93, 95–99.

Lyons, K., & Scrinis, G. (2009). Under the Regulatory Radar Nanotechnologies and their İmpacts for Rural Australia. In: Tracking Rural Change: Community, Policy and Technology in Australia, New Zealand and Europe, Merlan, F. (Eds.). Australian National University. E Press, Canberra, 151-171.

Malerba, M., & Cerana, R. (2016). Chitosan effects on plant systems. International Journal Molecular Sciences, 17, 996.

Martinelli, F., Scalenghe, R., Davino, S., Panno, S., Scuderi, G., Ruisi, P., Villa, P., Stroppiana, D., Boschetti, M., Goulart, L. R., Davis, C. E., & Dandekar, A. M. (2014) Advanced methods of plant disease detection:a review. Agronomy for Sustainable Development, 35,1–25.

Mishra, S., & Singh, H. (2015). Biosynthesized silver nanoparticles as a nanoweapon against phytopathogens: Exploring their scope and potential in agriculture. Appl. Microbiol. Biotechnol. 99, 1097–1107.

Mody, V. V., Cox, A., Shah, S., Singh, A., & Bevins, W. (2014). Parihar, H. Magnetic nanoparticle drug delivery systems for targeting tumor. Applied Nanoscience, 4, 385–392.

Park, H. J., Kim, S. H., Kim, H. J., & Choi, S. H. (2006) A new composition of nanosized silica-silver for control of various plant diseases. Plant Pathology Journal, 22,295–302.

Prasad, R., Bhattacharyya, A., & Nguyen, Q. D. (2017) Nanotechnology in sustainable agriculture: recent developments, challenges, and perspectives. Frontiers in Microbiology 8, 1014.

Prasad, R., Kumar, V., & Prasad, K. S. (2014) Nanotechnology in sustainable agriculture: present concerns and future aspects. African Journal of Biotechnology, 13(6), 705–713.

Prasad, R., Pandey, R., & Barman, I. (2016) Engineering tailored nanoparticles with microbes: quo vadis. WIREs Nanomed Nanobiotechnology, 8, 316–330.

Qing, S., Yongli, S., Yuehong, Z., Ting, Z., & Haoyan, S. Pesticide-conjugated polyacrylate nanoparticles: novel opportunities for improving the photostability of emamectin benzoate.

Rafique, M., Sadaf, I., Rafique, M. S., & Tahir, M. B (2017). A review on green synthesis of silver nanoparticles and their applications. Artif. Cells Nanomed. Biotechnology, 45, 1272–1291.

Rai, M., Yadav, A., & Gade, A. (2009). Silver nanoparticles as a new generation of antimicrobials. Biotechnology, 27, 76–83.

Russell, D., & Hugo, W. B. (1994). Antimicrobial activity and action of silver. Prog Med Chem 31:351–370 detection:a review. Agronomy for Sustainable Development, 35, 1–25.

Singh, S., Singh, B. K., Yadav, S. M., & Gupta, A. K. (2014). Applications of nanotechnology in agricultural and their role in disease management. Journal of Nanoscience and Nanotecnology.

Wainwright, M., Grayston, S. J., & deJong, P. (1986). Adsorption of insoluble compounds by mycelium of the fungus Mucor flavus. Enzyme and Microbial Technology, 8, 597–600.

Wen, Y. Z., Yuan, Y. L., Chen, H., Xu, D. M., Lin, K. D., & Liu, W. P. (2010). Effect of chitosan on the enantioselective bioavailability of the herbicide dichlorprop to Chlorella pyrenoidosa. Environment Science & Technology, 44(13), 4981–7.

Xu, L., Liu, Y., Bai, R., & C, Chen. (2010). Applications and toxicological issues surrounding nanotechnology in the food industry. Pure Applied Chemistry, 82, 349-372.

Yan, J., Huang, K., Wang, Y., & Liu, S. (2005). Study on anti-pollution nano-preparation of dimethomorph and its performance. Chinese Science Bulletin, 50, 108-112.

Yeo, S. Y., Lee, H. J., & Jeong, S. H. (2003). Preparation of nanocomposite fibers for permanent antibacterial effect. J. Mater. Sci. 38, 2143–2147.

7

Geo-spatial Technology for Pest and Diseases Management

***Mahesh Tripathi*[1] *and Himanshu Tripathi*[2]**

[1]*Department of Agricultural Engineering, Dr. K.S. Gill Akal College of Agriculture, Eternal University, Baru Sahib, Sirmour-173101 Himachal Pradesh*

[2]*Department of Agricultural Engineering, Faculty of Agricultural Sciences RSM College, Dhampur, Bijnor-246761 Uttar Pradesh*

Abstract

The agriculture sector faces significant challenges in mitigating the detrimental impacts of pests and diseases on crop yield and quality. Traditional methods of pest and disease management often fall short in providing timely and precise interventions, leading to substantial economic losses and environmental consequences. In recent years, the integration of geo-spatial technology has emerged as a promising approach to address these challenges by enabling more accurate monitoring, prediction, and management of pests and diseases.

The integration of geo-spatial technology in pest and disease management within agricultural practices represents a significant stride towards precision agriculture. This chapter delves into the innovative application of geo-spatial tools in identifying, monitoring, and mitigating pests and diseases, thereby optimizing crop yield and quality while minimizing environmental impact. Beginning with an overview of the challenges posed by pests and diseases in agriculture, the chapter highlights the limitations of traditional methods in effectively addressing these issues. It then explores how geo-spatial technology, encompassing Geographic Information Systems (GIS), remote sensing, and global positioning systems (GPS), revolutionizes pest and disease management by providing real-time, spatially explicit data. The discussion extends to the utilization of satellite imagery and aerial drones for early detection and monitoring of pest infestations and disease outbreaks. Through case studies and examples, the efficacy of geo-spatial technology in mapping pest distribution patterns, predicting outbreaks, and implementing targeted interventions is elucidated.

Furthermore, the chapter examines the role of geo-spatial technology in facilitating integrated pest management (IPM) strategies by integrating data on weather conditions, soil properties, and crop health. It explores how decision support systems (DSS) powered by geo-spatial analytics enable farmers and agricultural stakeholders to make informed decisions regarding pest and disease control measures, thereby optimizing resource allocation and minimizing pesticide usage.

Moreover, the chapter discusses the potential of geo-spatial technology in enhancing biosecurity measures and supporting quarantine protocols to prevent the spread of invasive pests and pathogens across regions.

In conclusion, the chapter underscores the transformative impact of geo-spatial technology on pest and disease management in agriculture, emphasizing its role in fostering sustainable farming practices, ensuring food security, and mitigating economic losses. It also outlines future research directions and challenges in harnessing the full potential of geo-spatial tools for the benefit of global agriculture.

Introduction

In recent years, the agricultural sector has witnessed a significant transformation with the integration of advanced technologies to tackle challenges such as pest and disease management. Among these technologies, geo-spatial technology has emerged as a powerful tool for monitoring, analyzing, and managing pest and disease outbreaks in agricultural fields. This chapter explores the applications of geo-spatial technology in pest and disease management, highlighting its benefits, challenges, and future prospects.

Plant stress is defined as a significant deviation from the optimal conditions for plant growth that could cause harmful effects when the limit of plants' ability to adjust is reached (Larcher 1995). Plant stress can affect almost every part of a plant, although typically one or few plant structures are influenced depending on the age and the source of stress. In case of biotic stress, the mechanism of damage by the pest largely infl- uences the physiological response of plants, which is in turn gets manifested into typical symptoms. Plants may respond to pest and disease stress in a number of ways, including leaf curling, wilting, chlorosis or necrosis of photosynthetic plant parts, stunted growth, or in some cases reduction in leaf area due to severe defoliation. Major types of pest damage mechanisms are classified as germination reduction, stand reduction, light stealing, assimilation rate reduction, assimilation sapping, tissue consumption and turgor reduction (Boote *et al.* 1983 ; Aggarwal *et al.* 2006) . While many of these responses are difficult to visually quantify with acceptable levels of accuracy, precision, and speed, these same plant responses will also affect the amount and quality of electromagnetic radiation reflected from plant canopies. Thus, remote sensing instruments that measure and record changes in electromagnetic radiation may provide a better means to objectively quantify disease stress than visual assessment methods. Furthermore, the effects of many pest/disease infestations are often not noticeable to the human eye, until it reaches an advanced stage when it becomes too late to control the outbreak. Remote sensing provides an alternative cost effective method to obtain detailed spatial information for entire crop fields at frequent intervals during the cropping season (Datt *et al.* 2006). Additionally, remote sensing can be used repeatedly to collect sample measurements non-destructively and non-invasively (Nilsson 1995; Nutter *et al.* 1990; Nutter and Litterell 1996). Pests and diseases cause serious economic losses in yield and quality of many cultivated crops, which is estimated at 14% of the total agricultural production (Oerke *et al.* 1994). Timely detection and assessment of their damage symptoms is very crucial. Traditionally, pest and disease assessment of crop plants is being done by a visual approach i.e., relying up on human eye and brain to assess their incidence. However the problem with the traditional approach is that they are often time consuming and labour intensive. Recent advances in the field of radiometry and other remote sensing technologies offer ample scope for exploiting these technologies towards developing an alternate means that can enhance or supplement the traditional approaches. Precise knowledge of areas where pest or disease activity has started would enable the farmer to apply just the right amounts of pesticides to the affected areas, thereby yielding both economic and environmental benefits (Datt *et al.* 2006). Though there are several distinct regions in the electromagnetic spectrum. In this review we restricted to the 'optical region' as this is the most extensively studied and widely used for remote sensing of biotic stresses in crop plants. Optical remote sensing makes use regions of visible, near infrared and short-wave infrared sensors.

Geospatial Technology (GST) refers to all emerging technologies, such as Remote Sensing (RS), Geographical Information Systems (GIS), and Global Positioning Systems (GPS), that help the user in the collection, analysis, and interpretation of spatial and temporal data. Geospatial technology has been employed in agriculture since the early 1990s. Farmers can now utilize a technology called the Global Navigation Satellite System (GNSS) to identify their agricultural equipment while also analyzing the spatial and temporal variation in soil, relief, and vegetation. This is possible because of the advancement of this system. This technology also improves the quality and accessibility of digital geographic data.

According to Mandal *et al*., 2013, Brazil utilized the geo-statistics method to generate a geographically differentiated nematode risk map in the field in order to address the issue of a loss in output of monoculture cotton of up to 40% which was brought on by an infestation of Rotylenchulus reniformis. In order to assess the risk of insects and diseases and the crop cultural requirements, ground-based weather, plant-stage measurements, and distant imaging were geo-referenced in GIS software.

Understanding Geo-spatial Technology

Geo-spatial technology encompasses a range of tools and techniques for gathering, analyzing, and visualizing geographic data. These include Geographic Information Systems (GIS), remote sensing, Global Positioning Systems (GPS), and spatial analysis software. By combining spatial data with attribute information, geo-spatial technology enables researchers and practitioners to gain insights into the spatial patterns of pests and diseases, identify high-risk areas, and implement targeted management strategies.

The term the Geospatial techniques are defined as the together with remote sensing, geographic information science, Global Positioning System (GPS), cartography, geovisualization, and spatial statistics are being used to capture, store, manipulate and analyze to understand complex situations to solve mysteries of the universe.

Benefits of Geospatial Technology

- Simplification of complex data to improve decision making.
- It helps in maintaining transparency in data for citizen access.
- Improved communication during crisis for better crisis management.
- It helps in managing natural resources.
- It helps the government in making better decisions.
- It helps in discovering precautions before planning development changes in a community.

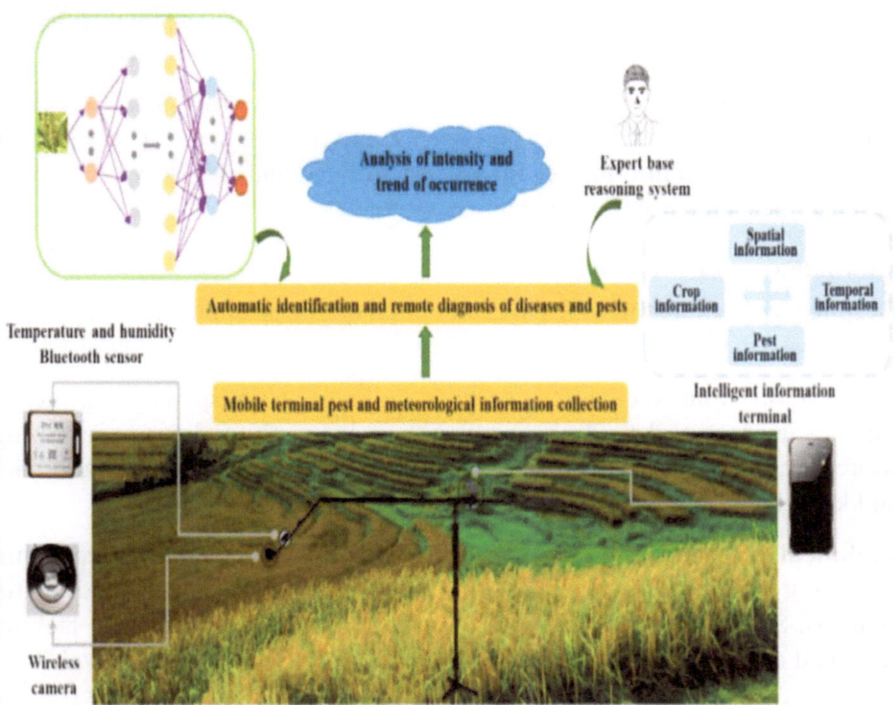

Fig. 1: Overview of intelligent plant protection monitoring system.
(Source: https://www.mdpi.com/2073-4395/12/5/1046)

Applications in Pest and Disease Management

Geo-spatial technology offers numerous applications in pest and disease management across various agricultural systems:

Early Detection and Monitoring: Remote sensing techniques, such as satellite imagery and aerial drones, can be used to detect early signs of pest and disease infestations. By analyzing spectral signatures and spatial patterns, researchers can identify areas of concern and track the spread of pests and diseases over time.

Risk Assessment and Mapping: GIS-based models can assess the risk of pest and disease outbreaks by integrating environmental factors, crop characteristics, and historical data. These models generate risk maps that highlight areas susceptible to infestations, enabling farmers to prioritize monitoring and control efforts.

Precision Pest Management: Geo-spatial technology facilitates precision agriculture practices by enabling targeted pest control interventions. GPS-equipped machinery can apply pesticides and other management practices with high spatial accuracy, minimizing input costs and environmental impact.

Decision Support Systems: Integrated geo-spatial platforms provide decision support tools for farmers, extension agents, and policymakers. These systems offer real-time monitoring, predictive modeling, and advisory services to optimize pest and disease management strategies.

Challenges and Limitations

Despite its potential, the widespread adoption of geo-spatial technology in pest and disease management faces several challenges:

Data Quality and Accessibility: Obtaining high-quality spatial data, such as satellite imagery and field observations, can be costly and time-consuming. Moreover, limited access to data and proprietary software hinders collaboration and knowledge sharing among stakeholders.

Technical Capacity and Infrastructure: Effective utilization of geo-spatial technology requires specialized skills in GIS, remote sensing, and spatial analysis. Smallholder farmers and resource-constrained regions may lack the technical capacity and infrastructure to implement these tools effectively.

Scale and Resolution: Balancing the scale and resolution of spatial data poses challenges in capturing fine-scale variations in pest and disease dynamics. High- resolution imagery may be expensive and computationally intensive, particularly for large agricultural landscapes.

Future Directions

To overcome these challenges and unlock the full potential of geo-spatial technology in pest and disease management, several avenues for future research and development are proposed:

Data Integration and Interoperability: Enhancing data interoperability and integration across platforms can improve accessibility and usability for stakeholders. Open data initiatives and standards-based protocols facilitate seamless exchange of spatial information.

Capacity Building and Training: Investing in capacity building programs and training initiatives can empower farmers and agricultural practitioners with the skills needed to leverage geo-spatial technology effectively. Extension services and public-private partnerships play a crucial role in disseminating knowledge and best practices.

Innovation in Sensor Technology: Continued innovation in sensor technology, including hyperspectral imaging and unmanned aerial vehicles (UAVs), holds promise for capturing detailed spatial information at various scales. Affordable and user-friendly sensor platforms expand the accessibility of geo-spatial tools to diverse user groups.

Conclusion

Geo-spatial technology offers tremendous opportunities for enhancing pest and disease management in agriculture. By harnessing the power of spatial data and analytical tools, stakeholders can make informed decisions, optimize resource allocation, and mitigate the impacts of pests and diseases on crop production. Addressing the challenges of data quality, technical capacity, and scalability is essential to realizing the full benefits of geo-spatial technology and ensuring its equitable adoption across agricultural systems.

References

Aggarwal PK, Kalra N, Chander S, Pathak H (2006) InfoCrop A generic simulation model for assessment of crop yields, losses due to pests and environmental impact of agro-ecosystems in tropical environments 1 Model description. Agric Syst 89:1–25

Boote KJ, Jones JW, Mishore JW, Berger RD (1983) Coupling pests to crop growth simulators to predict yield reduction. Phytopathology 73:1581–1587

Datt B, Apan A, Kelly R (2006) Early detection of exotic pests and diseases in Asian vegetables by imaging spectroscopy. Rural Industries Research and Development Corporation, Australia, RIRDC Publication No 05/170, pp 31

D. Mandal, P.P. Ghosh, M.K. Dasgupta Appropriate precision agriculture with site-specific cropping system management for marginal and small farmers Plant Sci Rev, 121 (2013), pp. 1-6

Larcher W (1995) Physiological plant ecology 3 rd edition. Springer, Berlin

8

Effect of Climate Change on Insect Pests and Food Security

Surendra Singh Shekhawat and Meenakshi Devi

SGT University, Gurugram-122 505, Haryana

Introduction

Agriculture plays a pivotal role in India's economy, it shares 18.2 % (2023-24) in country's GDP at current prices and employs 42.3 % of the Indian workforce, particularly in rural areas, making it one of the largest sources of livelihood (Anon., 2024). With 1.41 billion peoples, India is the world largest populous country and agriculture is crucial for ensuring food security. India's agricultural exports also significant in boosting its economy through generating foreign exchange of 53 billion dollars in 2023 and contributing around 11.8% to total exports (Statista, 2023). India is the largest producer of milk, millets, banana, sapota, cashew nuts, coconuts, tea, ginger, turmeric, black pepper and second largest producer of wheat, rice, sugar, groundnut and in the world (APEDA, 2024). The sector also provides raw materials for many industries, such as textiles and food processing. India achieved self-sufficiency in food grains after Green Revolution, using introduction of high-yielding varieties (HYVs) of crops, especially wheat and rice, chemical fertilizers, pesticides, and improved irrigation techniques, which dramatically increased crop productivity. The improved productivity also helped to increase farmers' incomes, raise living standards in rural areas, and support the development of rural infrastructure. Moreover, surplus food production allowed India to build buffer stocks, which could be used to manage food shortages or price fluctuations.

Challenges in Food Production

However, the Green Revolution helped secure India's food security, but it also brought several challenges, use of hybrid varieties, chemical fertilizers and pesticides resulted into more insect pest damage to the crops. Urbanization decreased the cultivated area and to feed increasing population, more numbers of crops are being produced on the same piece of land turned pest status and heavy load on the crops. To counter these insect-pests and to get more profit farmers started using pesticides injudiciously which also resulted into the several pest problems like; pest resistance, secondary pest outbreak, pest resurgence and negative impacts on both human health and the environment. In terms of human health, prolonged exposure to chemical pesticides has been linked to a range of health problems, including respiratory issues, skin conditions, neurological disorders, and even cancer.

From an environmental perspective, the use of chemical fertilizers and pesticides has led to soil pollution, reducing its natural fertility and harming the beneficial microorganisms that help maintain soil health.

Pesticides have also disrupted ecosystems by affecting non-target species, including pollinators like bees, which are vital for crop production. The decline in pollinator populations due to pesticide exposure poses a threat to biodiversity and agricultural productivity. Additionally, many pests have developed resistance to chemical pesticides over time, leading to a cycle of increased chemical use, further exacerbating environmental and health issues. This reliance on chemicals has created a significant ecological imbalance, affecting both the natural environment and the long-term sustainability of agricultural systems.

Economic Losses by Pests

In agriculture, major biotic limiting factor of the production includes; vertebrates (birds and rodents) and non-vertebrates (insect-pest, mites), diseases (fungi, bacteria, virus), plant parasitic nematodes and weeds. Accurate data of yield loss is not available but various organization and institute publish reports time to time. According to FAO, yield loss may reach up to 70% in the absence of control measures, among which weeds cause highest 30%, diseases – 23% and pests– 17%. The average loss done by the pests range 20 to 40 per cent with 290 billion dolalars, globally. Yield losses in wheat was estimated 7.7 per cent by weeds, 7.9% by animal pests and 12.6 per cent by pathogens. According to Dhaliwal and Arora () insect-pests, weeds and diseases can cause 10 to 30 percent loss including 25% in Rice, 5–10% in wheat, 20% in Sugarcane, 30 % in pulses, 35% on oilseeds, 50% in Cotton, annually. After harvest losses accounts for about 10 per cent of total stored grains of worth Rs 7000 crore due to poor storage infrastructures, insect, rodents and diseases. Severe incidence of animal pests and diseases also caused heavy economic loss in the past occurred in different parts of the world. Among them, many diseases and environmental changes have wiped out whole fields of crops; e.g., potato blight in the1800's, corn leaf blight – 1900's, cherry trees in northern Colorado in the 1950's.

Due to potato blight, the Irish Famine of 1846–1850 claimed up to a million lives due to disease and starvation, and it significantly altered Ireland's social and cultural landscape. Approximately 85 percent of US corn fields were planted with Texas cytoplasmic male sterile (Tcms) corn in 1970. Regretfully, this variety of corn exhibited a high susceptibility to a novel strain of the pathogenic fungus B. maydis race T. Extremely wet weather combined with TCMs corn's high susceptibility to B. maydis race T caused the pathogen to spread quickly, resulting in a devastating epidemic. Corn losses were disastrous, with some US regions experiencing losses of 50–100%.

The economic losses from southern corn leaf blight disease totaled about 1 billion dollars.

At least three quarters of a million people died during the great Bengal famine of 1943 AD in Bengal, a state in British-ruled India. The fungal pathogen Cochliobolus (Helminthosporium) caused brown spot disease in rice, which resulted in crop loss and the Bengal famine of 1943 AD.

When Hemileia vastatrix, the fungus that causes coffee rust, arrived in Ceylon in 1875, the island's coffee plantations covered almost 400,000 acres (160,000 hectares). The fungus was able to colonize the leaves until almost all of the trees were defoliated because there were no efficient chemical fungicides available to protect the foliage.

Banana production is seriously threatened by Fusarium wilt (FW), a disease caused by the soil- borne fungus Fusarium oxysporum f. sp. cubense (Foc). In the mid-twentieth century FW, also known as "Panama disease", wiped out the Gros Michel banana industry in Central America.

From 1907 to 1916, downy mildew reduced output of German vineyards by 33%, while significant periodic losses occurred in Italy in 1889, 1890, 1903, 1910, 1928, 1933 and 1934 (Müller, 1938). In 1915, 70% of the French grapevine production was destroyed by this pathogen. Introduced into Europe in 1875, Vitis vinifera was highly susceptible. Downy mildew spreads rapidly, destroying vineyards in France and most of Europe. In 1915, 70% of the French grapevine production was destroyed by this pathogen. In 1930, 20 million liters of wine were lost in France (Kolendenkova *et al.* 2022).

The most dangerous insect, locust is enemy of the maximum type of vegetation generally attacks in large groups known as "swarms" and produces havoc in the area where they attack. Recently in 2019, locust attack cause damage to the 1,79,584 ha of area of cultivation (PIB).

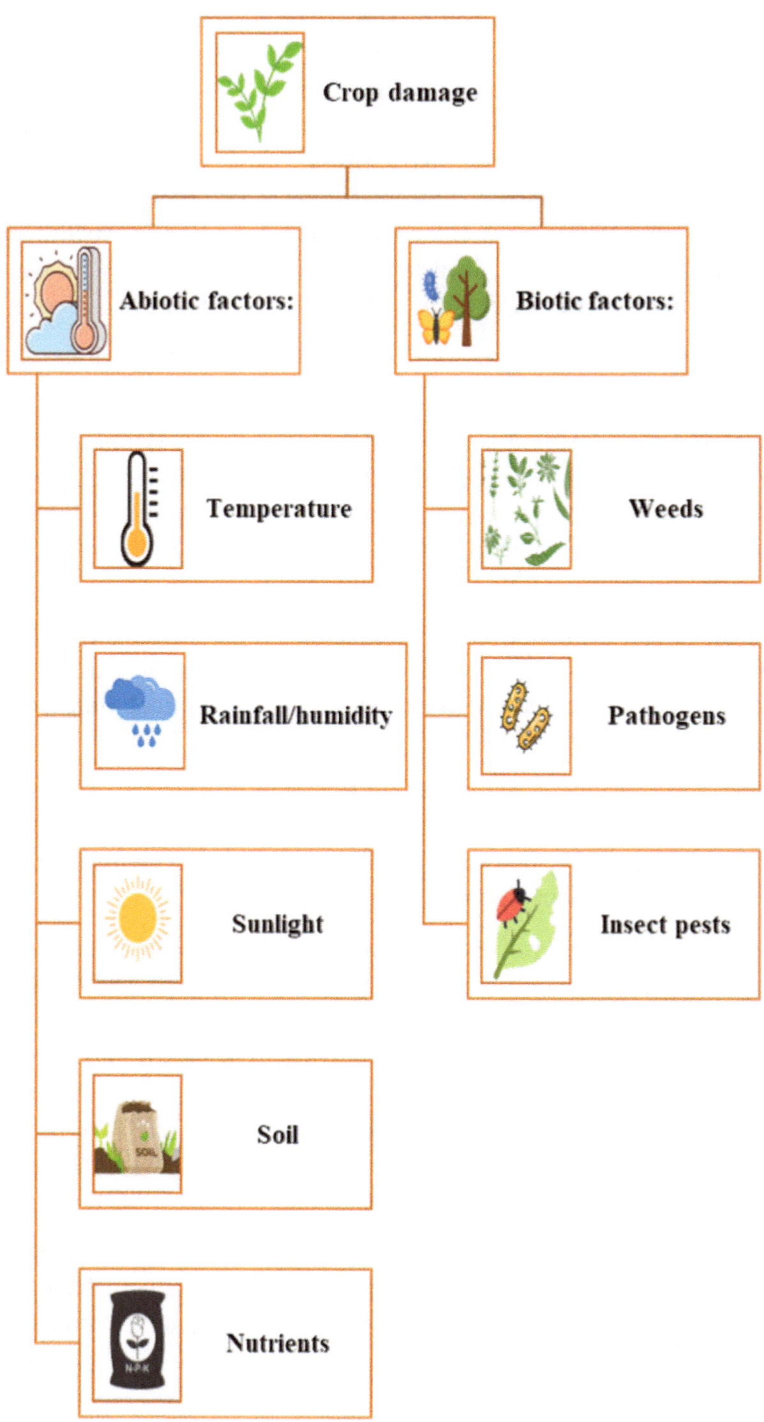
Crop damage
Abiotic factors:
Biotic factors:
Temperature
Weeds
Rainfall/humidity
Pathogens
Sunlight
Insect pests
Soil
Nutrients

Insect pest damage in agriculture is a major issue that affects crop productivity and food security worldwide. Pests feed on crops, causing direct damage to plant tissues such as leaves, stems, roots, and fruits, which results in reduced yields and quality. Some insects also act as vectors for plant diseases, further compounding the problem by spreading viruses, bacteria, or fungi that can devastate crops. The impact of insect pests varies by crop, season and region, but common pests like aphids, caterpillars, whiteflies, and locusts can destroy large areas of farmland in a short time. In some cases, pest outbreaks can cause widespread devastation, as seen with locust swarms that affect entire regions, leading to food shortages. Insect pest infestations can also force farmers to use chemical pesticides, which can have negative consequences for the environment and human health if not managed properly.

Table 1: Yield losses (%) due to weeds, pathogens and insect-pests in major crops

Crop	Weeds	Insect pest & Disease
Rice	13.8	10–30
Wheat	18.6	7.9–12.9
Maize	25.3	33–50
Mustard	21.4	31
Soybean	31.4	32
Sunflower	31.7	32
Pigeon pea	23.9	15–25
Groundnut	35.8	57
Chickpea	35.0	10–30
Sugarcane	21.9	20–30
Sorghum	25.1	30
Black gram	30.7	30
Green gram	30.8	30–50
Sesame	23.7	25
Cotton	17.9	20–40
Pearl millet	27.6	10–20

Effect of Climate Change on Insect-Pests

Climate change is influencing pest populations, with rising temperatures and altered weather patterns leading to changes in pest behavior, distribution, and reproduction rates. Warmer conditions often allow pests to thrive year-round, increasing the frequency and intensity of infestations. Insect pest damage also has long-term implications for agricultural sustainability, as repeated infestations can degrade soil health, increase production costs, and put added pressure on food systems.

Both agriculture and agricultural insect pests are significantly impacted by changes in the global climate. Climate change has an impact on agricultural crops and the pests that prey on them, both directly and indirectly. Indirect effects of climate change include those on the relationships that pests have with their surroundings and other insect species, including natural enemies, competitors, vectors, and mutualists. Direct effects are those on the reproduction, development, survival, and dispersal of pests (Prakash *et al.* 2014).

Because they are poikilothermic creatures, insects' body temperatures are influenced by the ambient temperature. Therefore, temperature is most likely the environmental factor influencing the behavior, distribution, development, and reproduction of insects. Thus, it is highly likely that the primary causes of climate change—increasing atmospheric CO_2, rising temperatures, and decreasing soil moisture—will have a substantial impact on insect pest population dynamics and, consequently, the proportion of crop losses (Kocmánková *et al.*, 2010; Fand *et al.*, 2012).

According to FAO, due to the creation of new ecological niches brought about by climate change, insect pests have more opportunity to proliferate, spread, and move between different geographic regions. The intricate physiological effects of increasing CO_2 and temperature can have a significant impact on how agricultural crops interact with insect pests (Hare *et al.*, 1992; Caulfield *et al.*, 1994; Roth *et al.*, 1995). As a result of the changing climate, farmers should prepare for new and severe pest problems in the years to come. Food security is threatened by the spread of crop pests across national and international borders, a worldwide issue that affects all nations and regions.

Changes in temperature greatly affect insect physiology, causing their metabolic rate to roughly double with every 10 °C rise (Dukes *et al.*, 2009). Many studies have demonstrated that higher temperatures can speed up insect eating, growth, and movement, impacting population dynamics through effects on reproduction, survival, time between generations, population size, and geographical distribution. Species that struggle to adjust and develop in response to rising temperatures tend to struggle to sustain their populations, while other species are able to flourish and reproduce quickly. Temperature is crucial for metabolism, metamorphosis, movement, and host availability, impacting the potential fluctuations in pest population and dynamics. Based on the way present-day insects are dispersed and act, it is possible to speculate that higher temperatures will result in more herbivorous activity. Based on how insect pests are spread and act, one could suggest that a rise in temperature would lead to more herbivory and alterations in the growth rate of insect populations. Insect populations in tropical zones may see a slower growth rate due to climate warming as temperatures are already ideal for pest development, while insects in temperate zones are likely to see a growth rate increase. The theory was verified by the authors through the evaluation of variations in pest population growth in the production of the top three grain crops worldwide (wheat, rice, and maize) across various climate change scenarios. Based on the research, warming temperatures will speed up the growth of pest populations for wheat, a crop typically cultivated in temperate regions. Tropical rice is expected to see a decrease in pest numbers, while maize grown in both tropical and temperate regions is also predicted to experience this trend. and for maize grown in both temperate and tropical regions, mixed responses to the growth of pest populations could be expected. Temperature greatly impacts the general functioning of the insects mostly who lives above the ground (Bale *et al.*, 2002). Increased temperature can disrupt the communication of alarm pheromone between the population and make them prone to predation. Likewise, high temperature and humidity also supports the population build up in whiteflies (Awmack *et al.*, 1997; Pathania *et al.*, 2020). Climate models has been predicted that the temperature will increase by 1.8 to 4 by the end of the century (Collins *et al.*, 2007). Studies showed that the increasing temperature will reduce the thermal constant of the insect development that will increase the severity of the infestation and pest abundance (Deutsch *et al.*, 2008, Lehmann *et al.*, 2020). Researches also showed that the global warming will also leads to the extinction of 15 to 37 % species by 2050. Species will expand their geographic range, invasiveness of alien spp., upsurge in insect transmitted diseases and increase in activity of overwintering population (Thomas *et al.*, 2004).

Response of Insect Pests to Increased CO_2

Increase in CO_2 levels can alter the population abundance and induce development rates. It also changed the feeding behavior of the insect on C_3 plant (Lincoln *et al.*, 1993). Increased CO_2 concentrations can alter chemical composition of the leaf and plant physiology by enhancing photosynthetic activity, resulting in improved growth and accumulation in the leaves can increase feeding by herbivory. Study showed that the increased CO_2 conditions can cause 57 per cent more damage in soybean by insect pests; leafhopper *Empoasca fabae,* Japanese beetle *Popilia japonica*, Mexican bean beetle *Epilachna varivestis* and Western corn rootworm *Diabrotica virgifera* (Hamilton *et al.*, 2005). Sap sucking insect pests like thrips, whiteflies and aphids exhibited extended population growth rates (Bezemer *et al.*, 1998). Stiling and Cornelissen (2007) showed mixed effects of CO_2 elevation on life parameters of the insects e.g. increased feeding in 17%, decrease in pest abundance in 22% and increase in development in 4%.

Effect of Rainfall Patterns

Changes in precipitation patterns are the key indicator of climate change. The increase in intensity and decrease in frequency promotes droughts and floods. Heavy rains are negatively correlated with survival of small insects and can wash out the eggs and other life stages of soft bodies insect like thrips, aphids, whiteflies, scales and mites. Soil dwelling insects and life forms can be affected by infrequent rainfall or flooding can kill these stages in the lack of oxygen (Pathak *et al.*, 2012). For example, increased summer rainfall is positively correlated with population of wireworms in grasslands, a potential threat to the potatoes, corn, sugar beet (Staley *et al.*, 2007). Climate change and international trade have a major impact on the geographic distribution of insect pests. In agricultural, introduction of insect pest species into the new region can cause heavy loss due to absence of its natural enemy. It results in increased pest complex and inputs in terms of money and work force to deal with it. Due to global warming, glaciers are melting and supporting vegetation which will be favoring insect also. likewise, the occurrence and distribution of European corn borer (*Ostrinia nubilalis*) has shifted more than 1000 km towards North (Porter *et al.*, 1991). Species are shifting their geographical distribution to seek survival and meet the criteria of optimum development threshold. Following agriculture on forest land also changing pest status of several species. the Diamondback moth (*Plutella xylostella* L.) has shifted its area by 800 Kms from Western Russia to Norwegian Islands (Coulson *et al.*, 2002). It has been also revealed that the pink bollworm (*Pectinophora gossypiella*), will expand its current range from southern Arizona into the cotton growing areas of Central California (Gutierrez *et al.*, 2006). Insects are cold-blooded animals and have a narrow range of preferred temperature for the development. Temperature beyond the optimum level leads to the diapause or inactivity of the insects. Extreme of temperature can result into the mortality, thus regarded as major limiting factor of the density. The low temperatures also increase mortality and reduce populations in the but upsurge in temperature may escape overwintering and induce activity. To deal with low temperature, insects generally utilize two modifications for survival, first they have physiological benefits or anti freezing compounds or within body and diapause, second they migrate to the plains to avoid winters. It has been noticed that low temperature prolonged the diapause period but increase in temperature shortens the diapause because temperature increase the metabolic rate of insects. Therefore, global warming would result in the increase in number of generations, high population build-up and prolonged activity. For example, increases in temperature resulted heavy loss by Corn earworm (*Helicoverpa zea*) and the Cotton bollworm (*Helicoverpa armigera*) due to prolonged activity (Diffenbaugh *et al.*, 2008).

Increased Number of Generations

Temperature increases development rate in the insects, that will result into the shorter life cycle and increased number of generation. It has been showed that a 2 °C uplift in temperature could result in 1-5 more generations per year (Yamamura and Kiritani, 1998). The most significant examples in this regard are aphids, which can be expected to produce four to five additional generations per year due to their low developmental threshold and short generation time. Aphids may, therefore, be particularly sensitive indicators of temperature changes. Higher temperatures during development have the advantageous effect of enabling species to become adults earlier, and they also shorten the time in the larval and nymphal stages (Bernays, 1997).

Effect on Natural Enemies

Climate change also have severe impacts on the population distribution, and seasonal abundance of pests and their natural enemies, and can influence the efficacy of biological control programs. Rise in temperature affects the development of the pest and parasitoid due to desynchronization or overlapping of generation (Hullé and Acier, 2010). Increase in CO_2 concentrations reduced the nutritional properties of alfalfa plants, which resulted in a decreased development time of the larvae of *Spodoptera exigua*. Due to this, the parasitoid, *Cotesia marginiventris* became extinct.

Increase in Insect-Vectored Plant Diseases

Due to geographic expansion and an increase in insect vector populations, plant diseases transmitted by insects may become more common as a result of global warming. The Hemiptera, an order of insects that feed on sap, are the primary carriers of plant viruses. The families of whiteflies (Aleyrodidae), leafhoppers (Cicadellidae), and aphids (Aphididae) are the principal vectors of viral diseases within this order. With over 275 different virus species that they can spread, aphids are by far the most common group of vectors among them. Furthermore, most aphid species can also spread certain plant viruses. While whiteflies are limited to warmer climates and are only found in temperate regions where crops are grown in greenhouses, aphids are important virus vectors in these regions (Canto *et al.*, 2009).

Aphids and whiteflies are particularly sensitive to responses to climate change due to their short development time and high reproductive capacity. Climate change may also have an impact on virus vectors' capacity for migration and long-distance dispersal. When aphids find favorable thermal conditions that propel them upward and expose them to horizontal translocation caused by atmospheric air movements, they can travel great distances. Aphids carried by incredibly persistent low-pressure winds from the southern Great Plains of North America to Minnesota's corn-growing regions have been linked to severe viral epidemics due to long-distance transport (Zeyen *et al.*, 1987).

Furthermore, it has been noted that higher temperatures in Northern Europe, particularly at the start of the growing season, have been linked to a higher incidence of viral diseases in potatoes. This is because aphids, the primary vectors of potato viruses, have been known to colonize potato plants earlier. The amount of inoculum and the timing of infection have a major impact on the severity of viral diseases. The overwintering of the virus's insect vectors and their (alternative) host plants affects the quantity of viral inoculum. Milder winters are predicted to result in higher survival rates for aphids, and warmer spring and summer temperatures promote their growth and reproduction. A higher incidence of viral disease transmission and spread is the ultimate result (Alonso-Prados *et al.*, 2003). The two most significant whitefly virus vectors are the greenhouse whitefly (*Trialeurodes vaporariorum*) and the silver leaf whitefly (*Bemisia tabaci*). High temperatures and moderate precipitation are typically favorable for *B. tabaci*, resulting in population growth. Situations where irrigation systems are installed and the climate is dry and hot are ideal for *B. tabaci*. Large populations can emerge in the summer due to their short generation time.

Pest Management in Climate Change

The process of putting current risk management techniques into practice and lowering the possible risk from climate change impacts can be seen as climate change adaptation. It is generally anticipated that pest infestations will become more erratic and expand geographically as a result of climate change. In addition, it is unclear how crop yields will be directly impacted by climate change and how insects and plants interact in ecosystems. Numerous biological, economic, and social factors will affect how adaptable agricultural production systems are. Local communities' financial, social, and physical resources will determine how easily they can modify their pest management strategies. The uncertainties surrounding the occurrence of both new and existing pests will rise as a result of climate change and the acceleration of global trade. Thus, it will be even more crucial to increase resilience to disruptions and climate change quickly (Barzman *et al.*, 2015). Adaptation strategies that could lessen the negative effects of current pests and lower the risks of introducing new diseases and pests have been identified. The most often mentioned tactics include using modeling prediction tools, monitoring insect pest populations and the climate, and modifying integrated pest management (IPM) practices.

IPM, by definition, includes pathogens, weeds, and harmful species of phytophagous animals (primarily insects and mites). Plant protection in sustainable agriculture places a strong emphasis on indirect or preventive measures, which must be fully utilized prior to the application of control or direct measures. The most up-to-date resources, such as forecasting techniques and scientifically verified thresholds, must be used to make decisions about the necessity of control measures. When indirect measures are unable to prevent economically intolerable losses, direct pest control tools should be considered a last resort (Boller *et al.*, 1999). The FAO advises using a dual approach that combines global and regional action with a major financial commitment to enhancing current early detection and control systems. To stop the spread, new crop species must be introduced, new agricultural techniques must be developed, and integrated pest management principles must be used. IPM techniques are primarily created by researchers and growers to reduce adverse environmental effects while increasing crop yields and financial gains. In order to improve heterogeneous agroecosystems that are resilient enough to withstand weather variability, many authors have addressed the issue of pest management in a novel environment with a changing climate and the need to reevaluate current preventive agricultural practises and IPM strategies. But in order to address the significant effects of global warming, it has been predicted in recent years that growers and researchers will need to modify many of these meticulously planned IPM strategies. Many IPM programs have focused on decisions based on extensive knowledge of how many insect pests can be tolerated before economic yield losses occur, also known as economic or intervention thresholds. In the past, IPM has developed in the field of pest management, where the application of set thresholds has produced positive outcomes. Intervention thresholds are crucial to IPM, but they're not always appropriate, sufficient, or feasible. Thresholds are not used when decision support systems are not appropriate or available (Barzman *et al.*, 2015). Crop advisors can adapt to climate change by having a thorough understanding of the ways in which the environment influences the development of plants and pests. Environmental elements like drought stress have an impact on crop protection guidelines. Drought stress reduces a crop's ability to withstand herbivorous insect stress, which can quickly lower the crop's economic threshold. Higher temperatures cause insects to develop more quickly, which accelerates population growth and causes crop damage earlier than is currently thought. In order to avoid unacceptably large yield losses, treatment thresholds based on the number of insects per plant must be lowered (Trumble and Butler, 2009). Pheromones and allelochemicals are two crucial ways that insects sense their surroundings. They contribute significantly to a number of IPM strategies, including monitoring, trapping, push-pull tactics, biological control, and mating disruption. The use of pheromones and allelochemicals in their current form is expected to become less effective as the climate warms and microclimates become more variable. To reduce their volatility under high temperature conditions, a synergist or other adjuvant may be necessary. Furthermore, some biopesticides that are based on bacteria, nematodes, fungi, or enthomopathogenic viruses are very sensitive to changes in their environment. Gaining more insight into how global warming affects the effectiveness of many synthetic pesticides, their durability in the environment, and the emergence of pest population resistance to specific insecticides is imperative. Consequently, it would seem necessary to take into account the application of effective biological control agents or the introduction of crop varieties resistant to insect pests that are obtained through genetic engineering or conventional genetic breeding. In response to climate change, existing pest management techniques like detection, prediction, physical control, chemical control, and biological control may need to be strengthened (Heeb *et al.*, 2019). Since many insect pests are transboundary, effective monitoring and risk assessment require a global management approach. It is necessary to have a worldwide system for exchanging data between regions, including critical information about diseases, invasive alien species, insects, and ecological conditions, including meteorological data. Therefore, it is important to improve cooperation between countries and regions, including national, regional, and global organizations. When dealing with invasive species, entry point monitoring and quick eradication—as demonstrated by the Early Warning and Rapid Response Program of the US Department of Agriculture and the Early Warning and Information System for IAS of the European and Mediterranean Plant Protection Organization (EPPO)—will remain crucial.Additionally, farmers can proactively implement specific pest prevention measures to decrease the incidence and increase of anticipated pest issues by keeping an eye on the climate, pests, and pest risk prediction data.

Conclusion

It is generally acknowledged that climate change has a significant impact on the cultivation of agricultural plants and the insect pests that are associated with them, even though there are still many unanswered questions about it. Some of the uncertainties surrounding various aspects of climate change that are pertinent to insect pests include small-scale climate variability, such as changes in relative humidity, temperature, atmospheric CO_2, and precipitation patterns. Insect species are expected to respond differently to global warming depending on where they live in the world due to the great diversity of these creatures, the plants that serve as their hosts, and the fluctuations in the global climate.

The impacts of climate change on insects are multifaceted, affecting their distribution, diversity, abundance, development, growth, and phenology, as well as favoring certain insects and hindering others. Furthermore, a general increase in the frequency of pest outbreaks involving a wider variety of insect pests is anticipated. It is likely that insects would spread geographically, particularly northward. The ability to produce more generations and an increased overwintering survival rate will lead to an increase in the abundance of certain pests. Plant diseases spread by insects will probably increase, and invasive pest species will probably spread more easily to new locations.

Reduced efficacy of natural enemies, or biological control agents, is another unfavorable effect of climate change that may arise and pose a significant challenge to future pest management initiatives. We run the danger of suffering large financial losses and jeopardizing the security of human food supplies if climate change-related factors create an environment that is conducive to pest infestation and crop damage. To address this issue, a proactive and scientific strategy will be needed. Planning and developing adaptation and mitigation strategies, such as modified integrated pest management (IPM) techniques, climate and pest monitoring, and the use of modeling tools, are therefore critically important.

References

Anonymous (2024). Data of PIB https://pib.gov.in/PressReleasePage.aspx?PRID=2034943#:~:text=Economic% 20Survey%20say s%20that%20the,country's%20GDP%20at%20current%20prices.

APEDA (2024). https://agriexchange.apeda.gov.in/India%20Production/Result_SearchProduct.aspx

Awmack, C., Woodcock, C., & Harrington, R. (1997). Climate change may increase vulnerability of aphids to natural enemies. Ecological Entomology, 22(3), 366-368.

Bale, J. S., Masters, G. J., Hodkinson, I. D., Awmack, C., Bezemer, T. M., Brown, V. K., & Whittaker, J. B. (2002). Herbivory in global climate change research: direct effects of rising temperature on insect herbivores. Global change biology, 8(1), 1-16.

Barzman, M., Bàrberi, P., Birch, A. N. E., Boonekamp, P., Dachbrodt-Saaydeh, S., Graf, B., & Sattin, M. (2015). Eight principles of integrated pest management. Agronomy for sustainable development, 35, 1199-1215.

Bernays, E. A. (1997). Feeding by lepidopteran larvae is dangerous. Ecological Entomology, 22(1), 121-123.

Bezemer, T. M., Jones, T. H., & Knight, K. J. (1998). Long-term effects of elevated CO 2 and temperature on populations of the peach potato aphid Myzus persicae and its parasitoid Aphidius matricariae. Oecologia, 116, 128-135.

Boller, E. F., El-Titi, A., Gendrier, J. P., Avilla, J., Jörg, E., & Malavolta, C. (1999). Integrated production. Principles and technical guidelines.

Canto, T., Aranda, M. A., & Fereres, A. (2009). Climate change effects on physiology and population processes of hosts and vectors that influence the spread of hemipteran-borne plant viruses. Global Change Biology, 15(8), 1884-1894.

Caulfield, F., & Bunce, J. A. (1994). Elevated atmospheric carbon dioxide concentration affects interactions between Spodoptera exigua (Lepidoptera: Noctuidae) larvae and two host plant species outdoors. Environmental Entomology, 23(4), 999-1005.

Collins, W., Colman, R., Haywood, J., Manning, M. R., & Mote, P. (2007). The physical science behind climate change. Scientific American, 297(2), 64-73.

Coulson, S. J., Hodkinson, I. D., Webb, N. R., Mikkola, K., Harrison, J. A., & Pedgley, D. E. (2002). Aerial colonization of high Arctic islands by invertebrates: the diamondback moth Plutella xylostella (Lepidoptera: Yponomeutidae) as a potential indicator species. Diversity and Distributions, 8(6), 327-334.

Crop loss https://www.cabi.org/projects/global-burden-of-crop-loss/

Deutsch, C. A., Tewksbury, J. J., Huey, R. B., Sheldon, K. S., Ghalambor, C. K., Haak, D. C., & Martin, P. R. (2008). Impacts of climate warming on terrestrial ectotherms across latitude. Proceedings of the National Academy of Sciences, 105(18), 6668-6672.

Deutsch, C. A., Tewksbury, J. J., Tigchelaar, M., Battisti, D. S., Merrill, S. C., Huey, R. B., & Naylor, R. L. (2018). Increase in crop losses to insect pests in a warming climate. Science, 361(6405), 916-919.

Diffenbaugh, N. S., Krupke, C. H., White, M. A., & Alexander, C. E. (2008). Global warming presents new challenges for maize pest management. Environmental Research Letters, 3(4), 044007.

Dukes, J. S., Pontius, J., Orwig, D., Garnas, J. R., Rodgers, V. L., Brazee, N., ... & Ayres, M. (2009). Responses of insect pests, pathogens, and invasive plant species to climate change in the forests of northeastern North America: what can we predict? Canadian journal of forest research, 39(2), 231-248.

Fand, B. B., Kamble, A. L., & Kumar, M. (2012). Will climate change pose serious threat to crop pest management: A critical review. International journal of scientific and Research publications, 2(11), 1-14.

FAO. Climate Related Transboundary Pests and Diseases. Available online: http://www.fao.org/3/a-ai785e.pdf

Gutierrez, A. P., D'Oultremont, T., Ellis, C. K., & Ponti, L. (2006). Climatic limits of pink bollworm in Arizona and California: effects of climate warming. Acta oecologica, 30(3), 353- 364.

Hamilton, J. G., Dermody, O., Aldea, M., Zangerl, A. R., Rogers, A., Berenbaum, M. R., & Delucia, E. H. (2005). Anthropogenic changes in tropospheric composition increase susceptibility of soybean to insect herbivory. Environmental entomology, 34(2), 479-485.

Hare, J. D. (1992). Effects of Plant Variation on Herbivore-Natural Enemy. Plant resistance to herbivores and pathogens: ecology, evolution, and genetics, 278.

Heeb, L., Jenner, E., & Cock, M. J. (2019). Climate-smart pest management: building resilience of farms and landscapes to changing pest threats. Journal of pest science, 92(3), 951-969.

https://www.agrivi.com/blog/yield-losses-due-to- pests/#:~:text=In%20the%20absence%20of%20crop,%25%20and%2017%25%2C%20respectiv ely.

Hullé, M., d'Acier, A. C., Bankhead-Dronnet, S., & Harrington, R. (2010). Aphids in the face of global changes. Comptes Rendus. Biologies, 333(6-7), 497-503.

Kocmánková, E., Trnka, M., Juroch, J., Dubrovský, M., Semerádová, D., Možný, M., & Žalud, Z. (2009). Impact of climate change on the occurrence and activity of harmful organisms. Plant Protection Science, 45(Special Issue).

Koledenkova K, Esmaeel Q, Jacquard C, Nowak J, Clément C, Ait Barka E. Plasmopara viticola the Causal Agent of Downy Mildew of Grapevine: From Its Taxonomy to Disease Management. Front Microbiol. 2022 May 11;13:889472. doi: 10.3389/fmicb.2022.889472. PMID: 35633680; PMCID: PMC9130769.

Lehmann, P., Ammunét, T., Barton, M., Battisti, A., Eigenbrode, S. D., Jepsen, J. U., & Björkman, C. (2020). Complex responses of global insect pests to climate warming. Frontiers in Ecology and the Environment, 18(3): 141-150.

Lincoln, D. E. (1993). The influence of plant carbon dioxide and nutrient supply on susceptibility to insect herbivores. Vegetatio, 104, 273-280.

Luis Alonso-Prados, J., Luis-Arteaga, M., Alvarez, J. M., Moriones, E., Batlle, A., Laviña, A., ... & Fraile, A. (2003). Epidemics of aphid-transmitted viruses in melon crops in Spain. European Journal of Plant Pathology, 109, 129-138.

Pathak, H., Aggarwal, P. K., & Singh, S. D. (2012). Climate change impact, adaptation and mitigation in agriculture: methodology for assessment and applications. Indian Agricultural Research Institute, New Delhi, 302, 19.

Pathania, M., Verma, A., Singh, M., Arora, P. K., & Kaur, N. (2020). Influence of abiotic factors on the infestation dynamics of whitefly, Bemisia tabaci (Gennadius 1889) in cotton and its management strategies in North-Western India. International journal of tropical insect science, 40, 969-981.

Porter, J. H., Parry, M. L., & Carter, T. R. (1991). The potential effects of climatic change on agricultural insect pests. Agricultural and Forest Meteorology, 57(1-3), 221-240.

Prakash, A., Rao, J., Mukherjee, A. K., Berliner, J., Pokhare, S. S., Adak, T., ... & Shashank, P. R. (2014). Climate change: impact on crop pests. Odisha, India: Applied Zoologists Research Association (AZRA), Central Rice Research Institute.

Roth, S. K., & Lindroth, R. L. (1995). Elevated atmospheric CO2: effects on phytochemistry, insect performance and insect-parasitoid interactions. Global Change Biology, 1(3), 173-182.

Staley, J. T., Hodgson, C. J., Mortimer, S. R., Morecroft, M. D., Masters, G. J., Brown, V. K., & Taylor, M. E. (2007). Effects of summer rainfall manipulations on the abundance and vertical distribution of herbivorous soil macro-invertebrates. European Journal of Soil Biology, 43(3), 189-198.

Statista (2024). https://www.statista.com/statistics/1300243/india-agricultural-export-value/#:~:text=In%20the%20fiscal%20year%202023,commodities%20exported%20in%20this% 20segment.

Stiling, P., & Cornelissen, T. (2007). How does elevated carbon dioxide (CO2) affect plant–herbivore interactions? A field experiment and meta-analysis of CO2-mediated changes on plant chemistry and herbivore performance. Global change biology, 13(9), 1823-1842.

Thomas, C. D., Cameron, A., Green, R. E., Bakkenes, M., Beaumont, L. J., Collingham, Y. C., & Williams, S. E. (2004). Extinction risk from climate change. Nature, 427(6970), 145-148.

Trumble, J., & Butler, C. (2009). Climate change will exacerbate California's insect pest problems. California Agriculture, 63(2).

Yamamura, K., & Kiritani, K. (1998). A simple method to estimate the potential increase in the number of generations under global warming in temperate zones. Applied Entomology and Zoology, 33(2), 289-298.

Zeyen, R. J., Stromberg, E. L., & Kuehnast, E. L. (1987). Long-range aphid transport hypothesis for maize dwarf mosaic virus: history and distribution in Minnesota, USA. Annals of applied biology, 111(2), 325-336.

9

Soil Health Management Innovative Tools for Plant Protection

***Ravi Dhanker*[1] *and Mukesh Kumar*[2]**

[1]*Department of Soil Science & Agricultural Chemistry, R.S.M. (PG) College Dhampur, Bijnor-246 761, Uttar Pradesh*
[2]*Natural Resource Management, College of Horticulture, S.D. Agricultural University, Jagudan-384 460, Gujarat*

Introduction

Soil health management plays a pivotal role in sustainable agriculture and directly influencing plant growth, crop productivity and environmental quality. Healthy soils are essential for nutrient cycling, water retention and supporting a diverse array of microorganisms that contribute to plant resilience. However, modern agricultural practices, including the excessive use of chemical fertilizers, monoculture cropping and improper irrigation, have degraded soil health, leading to nutrient imbalances, reduced organic matter and soil erosion. This has sparked a growing awareness of the need for effective soil health management strategies that not only maintain but enhance soil quality.

Innovative tools for soil health management are reshaping plant protection strategies by promoting holistic approaches that focus on maintaining and improving the biological, chemical and physical properties of the soil. These tools range from organic amendments and biofertilizers to advanced diagnostic technologies such as soil microbial profiling, remote sensing, and precision agriculture techniques. Together, they offer a comprehensive framework for addressing soil-related constraints and enhancing plant resistance to pests and diseases.

In this chapter, we explore the latest advancements in soil health management and their implications for plant protection. We discuss how integrating innovative tools can improve soil fertility, boost plant immunity and reduce dependency on chemical pesticides. By focusing on sustainable practices, this approach aligns with the goals of climate-smart agriculture and contributes to long-term agricultural productivity and environmental conservation. The chapter emphasizes the importance of adopting soil-centric strategies in modern agriculture to ensure resilient food systems and sustainable land use.

What is Soil Health?

Soil health refers to the condition of the soil in terms of its ability to sustain plant growth, maintain environmental quality and promote plant and animal health. It encompasses the physical, chemical and biological properties of the soil, which work together to support various ecosystem functions, including nutrient cycling, water filtration and carbon sequestration. When soil is healthy, it has a balanced structure, adequate nutrient availability, good water-holding capacity and a thriving microbial community that fosters beneficial interactions with plants.

The concept of soil health goes beyond just soil fertility, which typically focuses on the availability of nutrients for plant growth. It includes the soil's ability to function as a living system that can resist degradation, regenerate nutrients and adapt to changing environmental conditions.

Key Indicators of Soil Health

Important parameters of soil health provide valuable information about the soil's condition and its ability to perform essential functions. These indicators fall into three main categories: physical, chemical and biological, each contributing to a comprehensive understanding of soil health.

Physical Properties: Soil texture, structure, aeration, water retention, temperature and compaction levels.

Chemical Properties: Status of organic matter content, soil salinity, alkalinity, nutrient availability and toxic elements.

Biological Properties: Microbial diversity and enzyme activities.

Combining these physical, chemical and biological indicators provides a holistic view of soil health. Assessing multiple indicators allows for a more accurate evaluation, helping land managers implement practices that enhance soil quality and sustain agricultural productivity.

Innovative Tools for Soil Health Management

Advanced techniques and tools for soil health management provide cutting-edge methods and technologies to assess, maintain and improve soil quality. These tools develop understanding of soil properties, facilitate sustainable practices and help optimize plant growth while reducing environmental impact. They can be categorized into several groups as follows:

Soil Microbiome Enhancement

The soil microbiome, a complex community of microorganisms, plays a pivotal role in nutrient cycling and plant protection. There are two possible ways to enhance soil microbial diversity. The first one is the direct addition of microbial culture and the other is the enhancement of the inherent soil microbial population by providing a suitable environment for microbial growth. The direct improvement of soil microbial diversity was started with the use of biofertilizers (microbial inoculation having the capacity of nutrient acquisition/ fixation) for nitrogen fixation and other nutrient mobilization. The possible options for enhancing microbial diversity are as follows:

Microbial Profiling: Soil microbial profiling refers to the process of analyzing the composition, abundance and activity of microorganisms present in soil. These microorganisms include bacteria, fungi and other microscopic life forms that play a critical role in maintaining soil health, nutrient cycling, organic matter decomposition and plant growth. By understanding the microbial composition and activity in soils, it is possible to make informed decisions for sustainable land management, agricultural productivity and ecosystem conservation.

Native Microorganisms: Indigenous microbial communities naturally present in the soil, including bacteria, fungi, actinomycetes and other microscopic life forms. These microorganisms have adapted to the local soil and climatic conditions over time and play a vital role in maintaining soil fertility and ecosystem balance. The native microorganisms present in virgin soil that used in the preparation of formulations like *Jeevamrut* are key components in traditional and natural farming practices that emphasize sustainability and soil health.

Biofertilizers: Application of biofertilizers such as nitrogen-fixing bacteria (*e.g., Rhizobium, Azotobacter, Azospirillum, etc.*), phosphorus-solubilizing bacteria and Vesicular Arbuscular Mycorrhiza (VAM) can improve nutrient availability while reducing the need for chemical fertilizers.

Waste Decomposer: Waste decomposer is a microbial culture-based solution that accelerates the decomposition of organic matter. It contains a consortium of beneficial microorganisms that break down agricultural waste, crop residues and organic materials, converting them into valuable organic manure (Choudhary *et al.*, 2016).

Microbes for Soil Pollutants: Microbial bioremediation is the process of using specific microorganisms to break down or degrade agrochemical residues and soil pollutants. Certain microbes possess natural metabolic pathways that enable them to decompose harmful chemicals, detoxify pollutants and convert them into less toxic or harmless substances (Nayak *et al.*, 2018). Using microbial cultures to enhance the decomposition of agrochemical residues and soil pollutants is a promising approach for sustainable agriculture and environmental conservation.

Microbial Inoculants: Application of beneficial microbes like *Bacillus subtilis* and *Trichoderma* spp. helps to protect plants from root pathogens by outcompeting harmful organisms. These microbial interventions not only improve nutrient availability but also help in suppressing soil-borne diseases by increasing the beneficial microbial population.

Soil Organic Matter Management

Organic matter is the key factor to maintaining soil health and fertility. It enhances water retention, provides a habitat for microbes and plays a role in the structural stability of soils. The operation of following practices enhances the status of organic matter in soil.

Organic manures: The addition of organic matter through FYM, vermicompost, poultry manure, oil cakes and farm compost improve the organic carbon content of soil and introduces beneficial organisms that support plant growth.

Green Manuring: The incorporation of cover crops or legumes at flowering stage or leaves and soft branches of shrubs and trees in soil adds organic matter while fixing atmospheric nitrogen, improving soil structure and fertility.

Biochar: The use of biochar, a carbon-rich product obtained from organic material pyrolysis, enhances the soil's ability to retain nutrients and moisture while promoting microbial activity.

Crop Rotation and Cover Cropping

Crop rotation and cover cropping are traditional yet highly effective tools for improving soil health and plant protection.

Crop Rotation: Alternating between crops of different species breaks pest and disease cycles. For example, rotating legumes with cereals improves nitrogen content and reduces the risk of soil-borne diseases.

Cover Cropping: Planting cover crops in between main crop or sole like cowpea suppresses weeds, enhances organic matter and provides a habitat for beneficial insects and microorganisms.

These practices ensure that the soil remains fertile and less prone to diseases, reducing the need for chemical pesticides.

Mulching

Mulching is the practice of covering the soil surface around plants with a protective layer of material. This material, called mulch, can be organic (such as straw, leaves, grass clippings, or wood chips) or inorganic (like plastic sheeting, gravel or landscape fabric). Mulching offers several benefits:

Moisture Retention: Mulch helps retain soil moisture by reducing evaporation, which can be especially useful during hot and dry weather.

Weed Suppression: A thick layer of mulch prevents weed growth by blocking sunlight, reducing competition for nutrients and water.

Temperature Regulation: Mulch acts as an insulating layer, keeping the soil cooler in summer and warmer in winter.

Soil Health Improvement: Organic mulches decompose over time, adding organic matter to the soil, improving soil structure and increasing nutrient availability.

Erosion Control: Mulch protects the soil surface from erosion by wind and water.

Pest Control: Some types of mulch can help deter certain pests or reduce the spread of diseases by preventing soil from splashing onto plant leaves.

Conservation Tillage Practices

Conservation tillage reduces soil disturbance, which helps maintain soil structure and organic matter. Techniques like zero tillage and strip-tillage conserve soil moisture, reduce erosion and promote the buildup of organic matter (Kassam *et al.,* 2019).

No-till Farming: This technique leaves the soil undisturbed, preventing the destruction of beneficial microbial networks that help in nutrient cycling and disease suppression.

Reduced Tillage: Minimizing tillage operations preserves the microbial structure of the soil, helps reduce erosion and increases carbon sequestration, which benefits both plant health and the environment.

Biocontrol Agents

Innovative biocontrol agents are being utilized for plant protection by targeting specific soil-borne pathogens. These include:

Predatory Nematodes: Beneficial nematodes such as *Steinernema spp.* target and eliminate harmful pests in the soil without affecting non-target organisms. Entomopathogenic nematodes (EPNs) of the Steinernematidae and Heterorhabditidae families are widely used as biological control agents that represent a promising alternative to replace pesticides (Labaude and Griffin, 2018), because of their ability to parasitize insects, being able to identify, locate and infect a host and to kill it and they are safe to plants and other non-target organisms.

Fungal Antagonists: Fungi like *Trichoderma* spp. have antagonistic properties against many plant pathogens and promote root development by colonizing plant roots and production of antibiotics observed to suppressed the severity of the disease, particularly disease caused by soil-borne plant pathogens (Akhter *et al.*, 2015).

Pheromone Traps: While traditionally employed above ground for monitoring and managing pest populations, previous research has demonstrated that pheromone traps can also be effectively utilized to detect and control soil-dwelling insect populations. Studies have shown that adapting pheromone traps for use in soil environments can help monitor the activity and abundance of subterranean pests, offering an integrated approach to pest management by targeting insects during their soil-dwelling stages. This extension of pheromone trap technology provides a valuable tool for the early detection and reduction of soil-borne pest populations in agricultural systems.

Soil Testing and Precision Agriculture

Soil testing allows farmers to understand the exact needs of their soil, enabling the precise application of nutrients and amendments. The integration of technology into soil management has led to several innovations:

Soil Health Card: Based on the soil test report, it's easier to manage the nutrient requirements of a specific crop and apply them in a balanced manner, which helps maintain soil health. Healthy soil directly impacts plant health, enhancing the plant's natural defence mechanisms against diseases and insect pests.

Remote Sensing and GIS: By using geographic information systems (GIS) and satellite data, farmers can monitor soil moisture, nutrient levels and plant health. This data helps in making informed decisions on soil management and pest control.

Smart Irrigation Systems: These systems optimize water use based on soil moisture data, helping prevent waterlogging and root diseases.

Soil Amendments: Based on the results of soil tests, organic or inorganic amendments can be applied to maintain pH, improve soil structure and restore nutrient deficiencies.

Advanced Data Analytics and Machine Learning

Artificial Intelligence (AI) and Machine Learning Models: These can analyze large datasets from soil sensors, remote sensing and historical farm records to predict soil health trends and recommend management practices. AI-driven tools assist in optimizing soil inputs and improving decision-making in soil health management.

Decision Support Systems (DSS): DSS software integrates data from various sources to help farmers make informed decisions about soil management. They offer recommendations on selection of crop, date of sowing, crop rotation, fertilization, irrigation and other practices based on real-time soil health data.

These innovative tools empower farmers and land managers to adapt sustainable practices that enhance soil health, support plant growth and contribute to long-term agricultural productivity. By integrating these tools, modern agriculture can better manage soil resources, reduce environmental impact and adapt to changing climate conditions.

Soil Health and Plant Protection: A Symbiotic Relationship

Soil health and plant protection are deeply interconnected, forming a symbiotic relationship essential for sustainable agriculture and ecosystem stability. Healthy soils, rich in organic matter and microbial diversity, provide plants with essential nutrients and natural disease resistance. Beneficial soil microorganisms, such as bacteria and fungi, play a key role in nutrient cycling, organic matter decomposition and pathogen suppression. This thriving microbial community not only promotes stronger plant growth but also helps to outcompete harmful pathogens, reducing the risk of soil-borne diseases and fostering a resilient agricultural system.

In turn, plants contribute to soil vitality by releasing root exudates that feed soil microorganisms and by adding organic matter through leaf and root residues. These processes enhance microbial activity, improve soil structure and promote nutrient availability. As plant roots stabilize soil aggregates, they help maintain aeration and water retention, further supporting the soil ecosystem. By prioritizing practices that nurture this soil-plant symbiosis, such as using organic amendments, cover cropping and reduced tillage, farmers can enhance both soil health and plant protection, leading to more sustainable and productive farming systems.

Conclusion

Soil health management is essential for ensuring long-term plant protection and sustainability in agriculture. Innovative tools such as microbial inoculants, biofertilizers, organic matter management, precision agriculture and biocontrol agents are transforming how we approach plant protection by focusing on the health of the soil. By adopting these tools, we not only protect plants from pests and diseases but also contribute to the overall health of our ecosystems. The future of agriculture lies in creating resilient systems where healthy soil and robust plants work together to ensure productivity and environmental stewardship.

References

Akhter, W., Bhuiyan, M. K. A., Sultana, F., Hossain M. M. (2015). Integrated effect of microbial antagonist, organic amendment and fungicide in controlling seedling mortality (Rhizoctonia solani) and improving yield in pea (*Pisum sativum* L.). Comptes Rendus. Biologies, 338 (1): 21-28

Choudhary, M., Sharma, P. C., Jat, H. S., Nehra, V., McDonald, A. J. and Garg, N. (2016). Crop residue degradation by fungi isolated from conservation agriculture fields under rice-wheat system of North-west India. International Journal of Recycling of Organic Waste in Agriculture, 5: 349–360.

Kassam, A., Friedrich, T. and Derpsch, R. (2019). Global spread of conservation agriculture. International Journal of Environmental Studies, 76: 29–51.

Labaude, S. and Griffin, C. T. (2018). Transmission success of entomopathogenic nematodes used in pest control. Insects, 9: 1–20.

Nayak, S. K., Dash, B. and Baliyarsingh, B. (2018). 'Microbial remediation of persistent agro-chemicals by soil bacteria: An overview' in Microbial Biotechnology, Ed. Patra J., Das G. and Shin H. S. (Singapore: Springer), 13.11

Role of sustainable approaches in management of chickpea pod borer (*Helicoverpa*armigera Hubner) population in chickpea crop

10

Sustainable Management of Plant Diseases

Sachin Kumar Jain[1], Ajay Kumar[1], Abhishek Singh[1] and Dushyant Kumar[2]

[1]Amar Singh College, Lakhaoti, Bulandhshahr Lakhaoti-203407 Uttar Pradesh
[2]CCRD College, Muzaffarnagar, 251001, Uttar Pradesh

Abstract

Sustainable approaches to crop protection that reduce the environmental impact of chemical pesticides. This text emphasizes the efforts of global organizations like FAO to promote sustainable agriculture through various methods, include Eco-friendly alternatives for controlling plant diseases. Focusing on cultivar selection, soil fertility, and Integrated Pest Management (IPM), along with organic farming practices to reduce the reliance on chemical pesticides. The importance of border inspections to prevent the spread of quarantine pathogens. The role of forecasting models, precision farming, nanotechnology, and endotherapy in modern disease management. Use of biocontrol agents, systemic resistance inducers, gene-silencing technology, and plant nutrition to control diseases sustainably. Engaging multiple sectors to implement sustainable crop protection solutions. These efforts aim to create a more balanced and sustainable approach to agriculture, fostering resilience against plant diseases while minimizing the environmental footprint of farming practices.

Keywords: *Sustainable Management, Crop Protection, Diseases, IPM*

Introduction

The principles of sustainable agriculture have been defined for decades and are encapsulated in the Food and Agriculture Organization's (FAO) vision, which emphasizes an integrated system of plant and animal production that addresses various goals essential for long-term viability. Here's a comprehensive overview of sustainable agriculture and integrated pest management (IPM) principles, with a particular focus on disease management within the regulatory framework. The transition to sustainable agriculture is essential in addressing the growing food demands and environmental challenges of our time. The regulatory frameworks established by global organizations set a foundation for practices that support ecological health while ensuring economic viability for farmers. By integrating sustainable principles, effective pathogen management and innovative technologies, agriculture can become more resilient and environmentally friendly, ultimately benefiting society as a whole. According to the FAO, sustainable agriculture should:

1. **Satisfy Food and Fiber Needs**: Ensure adequate production to meet human dietary requirements.
2. **Enhance Environmental Quality**: Improve the health of natural resources, including soil, water, and biodiversity.
3. **Efficient Resource Use**: Maximize the use of nonrenewable resources while integrating natural biological cycles.
4. **Economic Viability**: Sustain the economic health of farming operations.
5. **Quality of Life**: Enhance the living standards of farmers and the broader community.

Diseases Management Methods

Preventive Measures

Selection of the Site and Cultivars

Overview of the factors influencing sustainable crop production in the context of climate change. The key points of this are following:

1. **Climatic Factors**: Climate plays a crucial role in determining the suitability of crops for specific regions. With climate change affecting weather patterns globally, farmers must adapt to new conditions, such as unexpected frost events in traditionally frost-free Mediterranean climates, which have impacted crop profitability.
2. **Adaptation to Climate Change**: Changing climate conditions necessitate adjustments in pathogen control strategies and agronomical techniques. New pathogens may adapt to these altered climates, requiring more robust disease prevention methods.
3. **Soil and Biotic Factors**: Beyond climate, **edaphic factors** (soil fertility, texture, and porosity) and **biotic factors** (presence of pollinators, pests, and pathogens) are important considerations for long-term crop success.

These factors become more critical as climate change introduces new challenges, including shifts in pest populations and pathogen prevalence.

4. **Cultivar Selection**: The choice of cultivars is key to achieving sustainable crop production. For **woody species**, the selection of the appropriate cultivar and rootstock is crucial, particularly when expanding to new cultivation areas. For **herbaceous crops**, surveys have shown that farmers prioritize eco-stability, grain yield performance, and steadiness when selecting cultivars, especially in response to changing climate conditions.

Healthy Seeds and Plant Material

The critical importance of maintaining a **healthy phytosanitary status** for seeds, tubers, plantlets, potted plants, and other propagative materials, especially given the increasing global trade in agricultural and forest commodities.

1. **Global Circulation and Phytosanitary Risks**: The significant rise in the global movement of plants and plant materials has heightened the risk of spreading pests and pathogens to new regions. These pathogens, once introduced, can infect local crops, potentially even on different continents.

2. **Factors Influencing Pathogen Establishment**: Several factors determine whether a phytopathogen becomes established in a new environment. These include the frequency of introductions, transmission rates, host density, climatic conditions, and the synchronization between host susceptibility and the pathogen's life cycle.

3. **Surveillance and Border Control**: Each country needs an **efficient surveillance system** at its borders to detect and intercept potential phytosanitary threats before they can take hold. This is especially crucial for countries lacking well-developed quarantine-based phytosanitary regulations.

4. **Preventive Measures for Seed Companies and Nurseries**: Companies and nurseries must adopt stringent preventive practices to reduce the risk of pathogen colonization. This includes implementing effective pathogen control during plant growth and disinfecting plant material prior to shipping.

5. **Farmer Vigilance**: Farmers should actively monitor their crops, especially during the early stages of growth, to detect and address potential diseases early, thereby minimizing the risk of crop loss.

Agronomical Techniques of Disease Management

1. **Agroecological Practices for Soil Fertility**: Crop rotation (especially with leguminous species) and planting cover crops between tree rows are effective long- term strategies for maintaining soil fertility. These methods also help control soil- borne diseases, such as those affecting potatoes (e.g., *Rhizoctonia solani* and *Streptomyces scabies*).

2. **Use of Biofertilizers and Soil Amendments**: The application of plant growth- promoting rhizobacteria (PGPR), biofertilizers, composts, mycorrhiza, biochars, and humic and fulvic acids enhances nutrient acquisition, improves soil texture, promotes plant growth, and induces systemic resistance to both biotic (e.g., pathogens) and abiotic (e.g., drought) stresses.

   For example, a biofertilizer containing mixed fungal and bacterial microflora protected bananas against *Fusarium wilt* after three years, and the fungus *Paecilomyces variotii* controlled *Fusarium wilt* in melons.

3. **Additions of Organic Matters**: While organic materials like animal manure can enhance soil fertility, they can also release antibiotics that disrupt native soil microflora, potentially harming crop growth.

4. **Balanced Crop Nutrition**: Ensuring balanced nutrition is an essential component of integrated crop protection programs, promoting healthier plants that are better able to resist diseases.

5. **Optimized Crop Protection Practices**: The calibration of sprayers, allowing the precise application of crop protection compounds to targeted plant parts (e.g., buds, leaves, canopy), reduces the spread of chemicals in the environment and minimizes water usage. This is a key aspect of reducing environmental impact in crop protection.

6. **Soil Solarization**: This technique, which involves covering the soil with plastic to trap solar heat.

7. **Organic Farming**: Organic farming greatly benefits from these preventive measures, particularly in controlling soil-borne pathogens without the use of synthetic chemicals, promoting healthier and more sustainable agricultural systems.

Current Control Strategies

Integrated Pest Management (IPM) as a sustainable strategy for controlling plant pathogens. IPM aims to reduce the use of chemical pesticides by integrating various plant protection methods. It encourages natural pest control mechanisms and supports the growth of healthy crops with minimal environmental disruption. The **FAO's IPM programme** in Asia, promotes sustainable agriculture by enhancing farmers' skills and awareness of agrochemical risks.

1. **Adaptability of IPM**: IPM approaches vary depending on the local agricultural environment, and each crop and region requires a tailored strategy. The combination of multiple control tactics often results in more effective and sustainable outcomes than a single-tactic approach.

2. **Technology in IPM**: Web-based platforms that offer real-time data on control thresholds, cultural practices, disease risks, and fungicide efficacy can help farmers, researchers, and advisors plan IPM strategies based on local environmental conditions.

3. **Systematic Knowledge Mapping**: Conducting studies to map out the control strategies currently applied to specific crops can help identify knowledge gaps. These maps have proven beneficial for applying IPM to large-scale arable crops like wheat, potatoes, and sugar beet in Sweden.

   Example

   - For **apple crops**: IPM strategies that combine disease monitoring, forecasting, eco-friendly fungicides, orchard sanitation, biological control, and insect control are comparable to conventional pest management.

   - The use of **resistant cultivars** enhances the effectiveness of biocontrol agents.

     For instance, *Bacillus mycoides* works well with sugar beet cultivars resistant to *Cercospora beticola*, and *Bacillus subtilis* effectively controls *Fusarium oxysporum* in chickpeas.

   - In **potato farming** in the Netherlands and Ireland, combining resistant potato cultivars with real-time monitoring of *Phytophthora infestans* populations reduced fungicide use by 80–90%.

4. **Seed Protection**: *Trichoderma gamsii*, applied to maize kernels, has shown effectiveness in reducing pink ear rot caused by *Fusarium verticillioides*.

5. **Organic Farming Synergy with IPM**: Organic farming principles, despite sometimes having a higher incidence of pathogens, can support effective biological control by increasing beneficial microorganisms in

crops. The synergistic relationship between IPM and organic farming promotes ecological intensification, improving environmental quality, farm viability, and both soil and human health.

Three Pillars of IPM and Organic Success

- **Disease forecasting models**
- **Biological control agents**
- **Environmentally friendly natural products**

Disease-Forecasting Models

Disease forecasting models are used to optimize the timing and method of applying agrochemicals or biocontrol agents, ultimately reducing pesticide use and protecting crops. These models integrate **climatic data** and the **pathogen's disease cycle** to assess when and how to apply treatments. They track the changes in an epidemic's components over time, influenced by factors like weather conditions, pathogen proliferation, and plant growth stages. This example demonstrates how integrating real-time environmental data into disease management systems can significantly improve the sustainability of agricultural practices by minimizing chemical inputs while maintaining crop health.

Biological Control

The passage delves into **biological control agents (BCAs)**, which are naturally occurring microorganisms that help control plant diseases by suppressing pathogen growth through various mechanisms. These mechanisms include competition for nutrients, antibiosis, induced resistance, predation, and the production of lytic enzymes. BCAs can be applied directly to crops as living organisms or through their metabolites.

Key Points on Biological Control Agents

1. **Mechanisms of Action**: BCAs work by competing for space and nutrients, producing antimicrobial compounds, or triggering the plant's natural defense systems. For example, *Pseudomonas chlororaphis* controls fungal pathogens, while *Trichoderma* spp. enhances plant resistance and controls several fungal diseases.
2. **Regulation of BCAs**: In countries like the United States and Canada, government agencies (EPA, ARLA) are responsible for ensuring the biosafety of BCAs. In Europe, BCAs must go through a detailed approval

process that involves multiple authorities, including the **European Food Safety Authority (EFSA)** and member state regulations.

3. **Nagoya Protocol**: The protocol restricts international exchange of biological materials, making it essential to focus on selecting **local biocontrol agents**. This is particularly relevant given the rise in organic food demand and sustainable agricultural policies.
4. **Selection and Effectiveness**: Selection of BCAs involves either screening for antimicrobial traits or using genetic/genomic approaches to identify desirable control traits. BCAs like *Trichoderma harzianum* have long-term effects in managing diseases like soybean root rot, while others like *Pseudomonas fluorescens* and *Bacillus subtilis* have been proven effective in a range of crop diseases.
5. **Durability**: The long-term efficacy of BCAs can vary, and environmental factors may reduce their effectiveness over time. For example, *Pseudomonas chlororaphis* showed decreased control of *Botrytis cinerea* after repeated treatments.
6. **Wide Application**: Several BCAs, such as *Trichoderma* spp., *Bacillus subtilis*, and *Pseudomonas fluorescens*, are versatile and used against various plant pathogens across crops. Yeast species like *Candida oleophila* and *Aureobasidium pullulans* are also used for postharvest disease control.
7. **Combining BCAs**: In some cases, a combination of BCAs can provide better disease control than individual agents, as seen in the combined use of *Bacillus mycoides* and *Pichia guillermondii* on strawberries. However, compatibility between agents should be carefully assessed to avoid reduced efficacy.

Challenges

Issues like genetic instability and limited shelf life, especially for pseudomonads, hinder the widespread commercial availability of some BCAs. Research into improving the shelf life and ensuring compatibility among agents can enhance their practical application.

Natural Products

Natural products with potential for controlling plant diseases are physiologically active chemicals derived from plants, microorganisms, or animals. These products serve as antimicrobials or inducers of systemic resistance and are known for being biodegradable, leaving minimal environmental residue.

Some natural products can even be the basis for synthetic pesticides, such as the fungicide strobilurin, derived from the fungus *Strobilurus tenacellus*. Common natural products used in disease control include chitosan, alkaloids, flavonoids, terpenes and phenolic compounds.

Natural Products for Disease Control

1. **Natural Product Composition**: Many of these products have evolved as **defense compounds** in their native organisms, showing potent bioactivities with specificity towards pathogens. This is particularly important as they often possess novel modes of action against pathogens resistant to conventional pesticides.

2. **Examples and Mechanisms**

   o **Harpin**, a protein elicitor derived from the bacterium *Erwinia amylovora*, triggers a plant's hypersensitive response, providing protection against pathogens. It has been effective in controlling blue mold on apples caused by *Penicillium expansum*.

   o **Reynoutria sachalinensis** extract is used to combat powdery mildew in cucumber and tomato as well as downy mildew in grapes by inducing plant defense mechanisms such as callose papillae formation and increasing levels of salicylic acid and caffeic acid.

   o **Chitosan**, a polymer derived from chitin, the primary component of fungal cell walls and crustacean exoskeletons, is commonly used in postharvest treatments to prevent fruit decay.

3. **Challenges with Antibiotics**: While antibiotics are naturally occurring products, many countries restrict their agricultural use due to concerns about the development of antibiotic resistance, which can affect both human and animal health.

4. **Emerging Technologies**: DNA sequencing technologies are being employed to identify genes or gene clusters that could help develop new natural products with unique modes of action. These innovations are crucial given the rise in **pesticide resistance** among crop pathogens.

5. **Environmental Impact**: One of the major benefits of natural products is their biodegradability. Unlike synthetic chemicals, they break down quickly in the environment, reducing pollution and minimizing long-term ecological impacts.

Sustainable Agriculture

We need to understanding the genomic structure, virulence factors, and epidemiology of pathogens for developing precise strategies to control biotic diseases in crops. For sustainable agriculture, we need to understand the following factors:

1. **Genomics and Epidemiology in Disease Control**: Knowledge of the pathogen's genomic structure and life cycle is essential for designing targeted biocontrol strategies. Biocontrol agents or curative compounds must be tested against different strains of the pathogen that represent the full population structure. This ensures the broad efficacy of the treatment.

2. **Timing of Biocontrol Applications**: Understanding the disease cycle of the pathogen is crucial for applying biocontrol agents or compounds at the right time—ideally before the pathogen colonizes the plant or multiplies internally. This timing is critical for effective disease prevention.

3. **Examples of Pathogen and Biocontrol Linkages**: Several studies demonstrate how the interaction between crop and pathogen epidemiology informs the selection of biocontrol agents and helps fine-tune the spread of active ingredients for disease management.

4. **Forecasting Disease Outbreaks**: Knowledge of pathogen life cycles also aids in the development of disease forecasting systems, allowing for early interventions before outbreaks occur.

5. **Sustainability and Social Acceptance**: Modern pathogen control must balance crop productivity with ecological and social considerations. Sustainable strategies should minimize environmental impact and be socially accepted by communities.

Developing Control Strategies

Modern Forecasting Models

Modern forecasting models for plant disease are essential tools in integrated pest management (IPM), providing vital information to growers. These models should encompass several key characteristics to enhance their effectiveness:

Characteristics of Modern Forecasting Models

1. Flexibility and Accuracy

- The models should adapt to various conditions, ensuring accurate predictions under different scenarios, such as varying climatic conditions and crop stages.

2. Differential Interactions

- The models must represent complex interactions among data inputs, including statistical relationships that reflect varying degrees of risk and infection dynamics.

3. Thresholds of Infection Risk

- It's crucial to define various thresholds of infection risks, considering potential mixed infections and the changing susceptibility of crops throughout the growing season.

4. Disease Dynamics Modeling

- Models should account for long-distance interactions and biological parameters, such as spore dispersal via wind and rainfall deposition. This holistic approach helps in understanding disease spread over large areas.

5. Precise Timing for Biopesticides Application

- They must provide accurate timing for applying biopesticides, integrating climatic and epidemiological data at different scales to optimize interventions.

Remote Sensing Technology

- Recent advancements in satellite monitoring have greatly enhanced the availability of spatial data. This technology allows for the incorporation of remote sensing data, such as leaf reflectance in the red band, which can indicate crop health and stress levels.
- Remote sensing can facilitate early detection of pre-symptomatic outbreaks and differentiate between symptoms caused by various pathogens, such as *Phytophthora infestans* and *Alternaria solani*.

Precision Farming

Precision farming offers innovative methods for managing plant diseases at the single-farm level, enabling real-time and site-specific predictions of pathogen outbreaks. The integration of technologies such as Geographic Information Systems (GIS) and on-site monitoring platforms allows farmers to assess pedo-climatic data continuously, providing timely alerts regarding potential disease outbreaks.

Key Components of Precision Farming for Disease Management

1. Geographic Information Systems (GIS)

- GIS plays a vital role in mapping and analyzing spatial data related to soil, weather, and crop health. By visualizing data in a geographical context, farmers can identify areas of the field that are more susceptible to disease, facilitating targeted interventions.

2. On-Site Monitoring Platforms

- These platforms continuously assess essential pedo-climatic parameters, including temperature, humidity, and soil moisture. By collecting real-time data, they can provide alerts to farmers about conditions conducive to pathogen development.

3. Phenomic Approach

- Utilizing non-invasive techniques to measure crop traits such as growth and performance enables early detection of disease-related changes that may not be visually apparent. This approach allows for precise timing of preventive or curative treatments, optimizing the application of agrochemicals.

4. Internet of Things (IoT) Monitoring

- IoT platforms incorporate artificial intelligence algorithms to simulate human decision-making processes. By analyzing data from various sources, these systems help determine the optimal timing for pathogen control measures, enhancing the overall effectiveness of interventions.

5. Agro-Weather Stations

- Installing agro-weather stations in the field allows for comprehensive data collection on climatic conditions. Measurements such as leaf wetness, soil temperature, water content, and solar radiation are crucial for detecting early signs of pathogen infection. This information helps farmers respond promptly to emerging threats.

6. Targeted Disease Management:

- One of the significant advantages of precision farming is its ability to manage disease based on localized occurrences within the farm. By identifying specific areas at risk, farmers can reduce pesticide usage and apply treatments more efficiently, minimizing environmental impact while effectively controlling diseases.

Benefits of Precision Farming for Pathogen Control

- **Real-Time Alerts**: Farmers receive immediate notifications about potential disease outbreaks, allowing for timely intervention.
- **Site-Specific Management**: Instead of blanket treatments, precision farming enables targeted actions, conserving resources and reducing chemical inputs.
- **Enhanced Decision-Making**: AI-driven insights support farmers in making informed choices regarding disease management, increasing the likelihood of successful outcomes.
- **Sustainability**: By reducing pesticide usage and optimizing treatment timing, precision farming contributes to more sustainable agricultural practices

Nanotechnology

Nanotechnology has the potential to significantly reduce the environmental impact of agrochemicals by enabling the targeted delivery of nanoparticles within plants. However, several critical factors need to be assessed to ensure the effective and safe application of nanomaterials in agriculture.

Considerations for Nanotechnology in Agriculture

1. Interactions with Plant Tissue

- Understanding how nanoparticles interact with plant tissues is essential. This involves studying the mechanisms of uptake, transport, and accumulation within the plant, which can affect both efficacy and safety.

2. Bioavailability and Durability

- The bioavailability of nanoparticles—their ability to be taken up and utilized by the plant—alongside their durability in the environment, are crucial parameters that influence their effectiveness as agrochemicals.

3. Ecotoxicological Risks

- Evaluating the potential ecotoxicological risks associated with nanoparticles is vital.

  This includes assessing their accumulation in food chains and their long-term effects on non-target organisms, including beneficial microbes and insects.

4. Cost-Effectiveness

- The economic viability of applying nanomaterials in agriculture must be considered.

  This includes not only the costs associated with the production of nanoparticles but also the potential savings from reduced agrochemical use.

Nanoparticles with Potential Applications

Several types of nanoparticles have shown promise as potential nanopesticides, including:

- **Copper, Zinc, and Manganese Nanoparticles**: Used for seed coatings and foliar applications, improving plant growth and disease resistance.
- **Silver Nanoparticles**: Known for their antimicrobial properties and have been shown to be effective against various plant pathogens.
- **Chitosan Nanoparticles**: Derived from biodegradable polymers, these nanoparticles exhibit good antimicrobial activity while being non-toxic to humans and animals.
- **Carbon and Magnesium Nanoparticles**: Other promising candidates for various agricultural applications.

Efficacy in Pathogen Control

Research indicates that certain nanoparticles exhibit significant antifungal activities against various plant pathogens:

- **Copper-Based Nanoparticles**: Effective against pathogens like *Fusarium oxysporum* in watermelons and *Verticillium dahliae* in eggplants.
- **Nanoparticles from Chaetomium cochlioides**: Demonstrated a reduction in rice blast severity caused by *Magnaporthe oryzae*.
- Studies show potential in reducing pathogenic activities in crops such as tomatoes, which face threats from *Fusarium* and *Botrytis* species.

Endotherapy

The technique of **trunk injection** or **trunk infusion**, also known as **endotherapy**, presents a promising method for delivering protective products directly into the xylem tissue of trees. This approach has the potential to significantly reduce the dispersal of agrochemicals into the environment, mitigate toxicity risks for farmers, and lower the overall costs associated with protective treatments.

Efficacy in Disease Control

- **Fire Blight Control**: In apple trees infected with *Erwinia amylovora*, trunk injections of products like **acibenzolar-S-methyl** and **potassium phosphite** have shown significant reductions in both blossom and shoot blight. These treatments also induced the expression of defense-related proteins.
- **Olive Quick Decline Syndrome**: Trunk injections of a biocomplex containing **zinc**, **copper**, and **citric acid** helped combat *Xylella fastidiosa* subsp. **pauca**. This treatment not only increased plant defense-related metabolites such as **oleuropein** and **polyphenols** but also reduced disease-related metabolites like **quinic acid** and **mannitol**.

Advantages of Endotherapy

- **Targeted Delivery**: This method allows for the precise delivery of active ingredients directly to the plant's vascular system, enhancing their efficacy and reducing the need for widespread application.
- **Reduced Environmental Impact**: By limiting the dispersion of chemicals into the environment, endotherapy presents a more sustainable approach to plant protection.
- **Farmer Safety**: With lower exposure to potentially toxic agrochemicals, farmer safety is improved, minimizing health risks associated with traditional pesticide application methods.

Biocontrol Agents

The potential of biocontrol agents in sustainable plant protection is promising but currently hampered by various challenges. Here are the key factors affecting their effective use, along with opportunities for development and application:

1. Fragmentation of Biocontrol Sub-disciplines

- The lack of integration between different areas of biocontrol research can hinder the identification and application of effective agents. Collaboration among various fields is essential for maximizing the benefits of biocontrol.

2. Bureaucratic and Regulatory Barriers

- The regulatory processes for the approval and registration of biocontrol agents can be cumbersome and expensive. High costs associated with registration and a lengthy approval process deter many companies from developing and marketing biocontrol products.

3. Economic Viability

- Farmers often consider cost-effectiveness when selecting biocontrol options.

  Companies must focus on improving formulation, storage, and delivery methods to reduce costs.

4. Public Engagement

- Insufficient communication about the benefits of biocontrol to stakeholders, including growers and policymakers, limits the acceptance and use of biocontrol strategies. Increased outreach is necessary to highlight economic and ecological advantages.

5. Biosafety and Environmental Impact Monitoring

- Any microorganisms released into the environment must be carefully monitored for potential negative impacts. Well-designed ecological monitoring programs can provide crucial data for regulators.

Successful Characteristics of Biocontrol Agents

1. **Niche Adaptability**: Agents should thrive in the specific environments of their target crops.
2. **Nutrient Competition**: Effective biocontrol agents compete successfully for nutrients with phytopathogens.
3. **Colonization Ability**: The ability to establish and persist in the target environment is critical for success.
4. **Antimicrobial Activity**: Preliminary screenings for antimicrobial compounds and antagonistic activities should be conducted on a wide range of isolates to identify promising candidates for further testing.

Lactic Acid Bacteria as Biocontrol Agents

- Lactic acid bacteria (LAB), typically recognized for their role in food preservation, also exhibit biocontrol potential against plant pathogens. For example, *Lactobacillus plantarum* has shown broad-spectrum antagonism against several bacterial phytopathogens, including:

- *Pseudomonas syringae* pv. *actinidiae* (kiwifruit bacterial canker)
- *Xanthomonas arboricola* pv. *pruni* (bacterial spot of stone fruit)
- *Xanthomonas fragariae* (angular leaf spot of strawberry).

Natural Products

Natural products derived from plants, animals, and microbial metabolites are gaining attention as effective biocidal agents in sustainable agriculture. Natural products such as essential oils, chitosan, and specific plant extracts offer sustainable alternatives for managing plant diseases. Their diverse mechanisms of action and effectiveness against various pathogens highlight their potential role in integrated pest management (IPM) strategies. Continued research into enhancing their stability, reducing costs, and overcoming regulatory barriers will be crucial for their broader adoption in agriculture.

Essential Oils

1. **Biocidal Activities**

    - Essential oils, composed of terpenes, hydrocarbons, alcohols, and phenols, have demonstrated effective biocidal activities against various phytopathogens, including fungi, oomycetes, and bacteria. They are known for their ability to avoid inducing pathogen resistance.
    - Oils from species such as *Mentha arvensis*, *Citrus x sinensis*, *Juniperus mexicana*, and *Cinnamomum zeylanicum* are registered in Europe for agricultural use.

2. **Field Applications**

    - Essential oils derived from *Zataria multiflora* effectively reduced cabbage black rot caused by *Xanthomonas campestris* pv. *campestris*.
    - Clove oil from *Eugenia caryophyllata* successfully controlled pomegranate bacterial blight caused by *Xanthomonas axonopodis* pv. *punicae*.
    - Oils from cumin, basil, and geranium showed effectiveness against cumin root rot caused by *Fusarium* species when applied as seed protectants or in the soil.

Chitosan: Chitosan has shown promise both as a plant defense inducer and in directly controlling diseases. Its film-forming ability protects plant surfaces, preventing pathogen colonization. Crops such as grapes, strawberries, apples, and citrus benefit from chitosan treatments, especially for postharvest decay.

For instance, it has effectively reduced *Pseudomonas syringae* pv. *actinidiae*, the agent of kiwifruit bacterial canker.

Plant Extracts

- *Equisetum arvense* (common horsetail) demonstrated control comparable to copper compounds against *Phytophthora infestans*, *Puccinia triticina*, and *Fusarium graminearum*.
- Pomegranate peel extract significantly reduced the severity of diseases caused by *Pseudomonas syringae* pv. *tomato* and *Fusarium oxysporum* f. sp. *lycopersici*. It also showed efficacy in postharvest control of *Botrytis cinerea* and *Penicillium* species.
- Extracts from *Camellia sinensis* displayed antibacterial activities against *Pseudomonas syringae* pv. *actinidiae*.

Nutrition Management

Nutrients play crucial roles not only in plant growth and development but also in plant pathogenesis and disease management. The strategic application of nutrients can be a viable approach for managing plant diseases in sustainable agriculture. Understanding the specific roles and interactions of different macronutrients and micronutrients can lead to more effective disease management practices, reducing the reliance on chemical pesticides and enhancing overall plant health. Further research into the precise effects of these nutrients in different pathosystems will be critical for optimizing their use in agricultural practices.

Nutrients and Disease Control

1. Balanced Nutrition

- Preventive balanced nutrition combined with rational agronomic techniques can be more effective and environmentally friendly compared to conventional pesticide applications. Each pathosystem (the interaction between a specific pathogen and crop in a given environment) has unique characteristics that should be studied to understand nutrient-disease relationships throughout the growing season.

2. Macronutrients

- **Nitrogen:** Can have dual effects on disease severity. It tends to increase disease severity with obligate pathogens but may reduce it with facultative pathogens.

- **Phosphorus:** Its role in disease control is inconsistent; however, potassium phosphate has shown effectiveness in reducing diseases like barley powdery mildew.
- **Potassium:** Generally has positive effects in reducing disease severity. Potassium phosphite, when applied within regulatory limits, can also help manage diseases like *Alternaria solani* on potatoes.
- **Calcium:** Particularly effective for post-harvest treatments against gray mold caused by *Botrytis cinerea.*
- **Magnesium:** Results regarding its efficacy in disease control are inconsistent and warrant further research.

3. Micronutrients

- **Copper:** Historically used as a fungicide and bactericide, but its excessive use can lead to soil accumulation, posing risks to microbial communities.
- **Zinc and Manganese:** Both micronutrients exhibit beneficial effects against plant diseases. For example, applying zinc-rich manure before planting winter wheat significantly reduced the severity of spring blight caused by *Rhizoctonia cerealis*. Manganese uptake is higher in paddy-grown rice, providing better resistance against rice blast (*Magnaporthe oryzae*).
- **Iron:** While it can enhance pathogen virulence, adding beneficial microorganisms to the rhizosphere can limit soilborne pathogens by activating siderophores to reduce soil iron content.

Systemic Resistance Inducers

Inducing systemic resistance in plants is a promising strategy for integrating pest management (IPM) approaches, leveraging the plant's natural defense mechanisms against pathogens. Here's an overview of the mechanisms, potential applications, and examples of inducers that enhance plant resistance:

Mechanisms of Induced Resistance

1. Systemic Acquired Resistance (SAR)

- Mediated by salicylic acid or its precursors, SAR is a defense response that helps plants resist a wide range of pathogens after an initial attack.

2. Induced Systemic Resistance (ISR)

- Triggered by plant growth-promoting rhizobacteria (PGPR), ISR involves the modulation of jasmonate and ethylene signaling pathways, leading to enhanced resistance against pathogens.

Commercially Available Resistance Inducers

- **Acibenzolar-S-methyl**: A widely used synthetic compound that induces resistance against various pathogens across different crops. It has been shown to protect mangoes from postharvest decay caused by *Colletotrichum gloeosporioides*, and it effectively protects fava beans from *Uromyces viciae-fabae* and *Ascochyta rabiei*. In combination with copper treatments, it reduced kiwifruit bacterial canker caused by *Pseudomonas syringae*.
- **Harpin**: A protein from plant pathogenic bacteria that stimulates plant defenses.
- **Chitosan**: A biopolymer with various plant protection applications.
- **Reynoutria sachalinensis** and **Saccharomyces cerevisiae** extracts: Both are known for their resistance-inducing properties.

Gene Silencing

RNA interference (RNAi) technology is emerging as a revolutionary strategy for controlling plant diseases by targeting the genes responsible for pathogenicity and virulence in pathogens. This method represents a shift towards sustainable agricultural practices by using plant-induced silencing to combat pathogens, including viruses. Here's an overview of how RNAi works, its applications, successes and challenges:

Mechanism of RNA Interference

1. Host-Induced Gene Silencing

- RNAi involves the introduction of small interfering RNAs (siRNAs) or exogenous non-coding RNAs into plants. These molecules can silence specific pathogen genes critical for their survival and pathogenicity.
- This process can occur via direct uptake from the environment or through engineered plants that produce these silencing RNAs.

2. Cross-Kingdom RNA Interference

- This innovative approach allows for the silencing of pathogen genes using host- derived RNA, effectively employing the plant's own defense mechanisms against invading pathogens.

Applications of RNAi in Plant Disease Control

1. Fusarium graminearum

- A significant reduction in pathogenicity was achieved on barley leaves by applying a spray containing double-stranded RNA (dsRNA) targeting *FgAGO* and *FgDCL* genes, which are integral to the fungus's RNA interference machinery. .

2. Phytophthora infestans

- The causal agent of potato late blight, *Phytophthora infestans*, showed reduced pathogenicity through gene silencing targeting the *hp-PiGPB1* gene, responsible for coding the G protein β-subunit.

3. Botrytis cinerea

- This fungus, responsible for gray mold on many crops, can be managed by spraying exogenous RNAs directed at the *DCL1* and *DCL2* genes, which are involved in pathogenicity. This approach has demonstrated significant disease reduction in various crops such as tomato, lettuce, onion, strawberry, and grapevine.

Conclusion

The effective control of plant diseases in horticulture is crucial not only for food production but also for the conservation of agroecosystems. As the global population grows and climate change poses significant challenges, the urgency for sustainable agricultural practices has never been greater. Here's an overview of the key points regarding the relationship between plant disease management, agroecosystem conservation and sustainable practices.

References

Abou-Shady, A.; Siddique, M.S.; Yu, W. A Critical Review of Recent Progress in Global Water Reuse during 2019–2021 and Perspectives to Overcome Future Water Crisis. Environments 2023, 10, 159.

Ahmed, F.; Raffi, M.; Ismail, M.; Juraimi, A.; Rahim, H.; Asfaliza, R.; Latif, M. Waterlogging Tolerance of Crops: Breeding, Mechanism of Tolerance, Molecular Approaches, and Future Prospects. Biomed Res. Int. 2012, 2013, 1–10.

Ali, O.; Cheddadi, I.; Landrein, B.; Long, Y. Revisiting the relationship between turgor pressure and plant cell growth. New Phytol. 2023, 238, 62–69.

Almaguer-Vargas, G.; Alcántar-González, G.; Osuna-Ceja, M. Production of hydroponic strawberry (Fragaria x ananassa Duch.) in response to electrical conductivity of the nutrient solution. Agrociencia 2008, 42, 641–652.

Antolini, F.; Tate, E.; Dalzell, B.; Young, N.; Johnson, K.; Hawthorne, P. Flood Risk Reduction from Agricultural Best Management Practices. J. Am. Water Resour. Assoc. 2019, 56, 161–179.

Ayars, J.; Phene, C.; Hutmacher, R.; Davis, K.; Schoneman, R.; Vail, S.; Mead, R. Subsurface drip irrigation of row crops: A review of 15 years of research at the Water Management Research Laboratory. Agric. Water Manag. 1999, 42, 1–27.

Ayars, J.E.; Christen, E.W.; Hornbuckle, J. Controlled drainage for improved water management in arid regions irrigated agriculture. Agric. Water Manag. 2006, 86, 128– 139.

Brown, M.; Bondurant, J.; Brockway, C. Subsurface trickle irrigation management with multiple cropping. Trans. ASAE 1981, 24, 1482–1489.

Bwambale, E.; Abagale, F.K.; Anornu, G.K. Smart irrigation monitoring and control strategies for improving water use efficiency in precision agriculture: A review. Agric. Water Manag. 2022, 260, 107324

Carrubba, A.; Militello, M. Growing peppermint (Mentha piperita L.) in hydroponics: A review. J. Med. Plants Res. 2013, 7, 3021–3029.

Chen, Y. The design of intelligent drip irrigation network control system. In Proceedings of the 2011 International Conference on Internet Technology and Applications, Wuhan, China, 16–18 August 2011; pp. 1–3.

Cordeiro, S.; Ferrario, F.; Pereira, H.Z.; Ferreira, F.; Matos, J.S. Water Reuse, a Sustainable Alternative in the Context of Water Scarcity and Climate Change in the Lisbon Metropolitan Area. Sustainability 2023, 15, 12578.

Coskun, Ö.F. The Effect of Grafting on Morphological, Physiological and Molecular Changes Induced by Drought Stress in Cucumber. Sustainability 2023, 15, 875.

Dalal, A.; Bourstein, R.; Haish, N.; Shenhar, I.; Wallach, R.; Moshelion, M. Dynamic Physiological Phenotyping of Drought-Stressed Pepper Plants Treated With "Productivity-Enhancing" and "Survivability-Enhancing" Biostimulants. Front. Plant Sci. 2019, 10, 905.

Drury, C.F.; Tan, C.S.; Reynolds, W.D.; Welacky, T.W.; Calder, W.; McLaughlin, N.B. Reducing nitrate loss in tile drainage water with cover crops and water-table management systems. J. Environ. Qual. 2009, 38, 1193–1204.

Emongor, V.E.; Ramolemana, G.M. Treated sewage effluent (water) potential to be used for horticultural production in Botswana. Phys. Chem. Earth 2004, 29, 1101– 1108.

Feset, S.E.; Strock, J.S.; Sands, G.R.; Birr, A.S. Controlled drainage to improve edge- of-field water quality in southwest Minnesota, USA. In Proceedings of the 9th International Drainage Symposium Held Jointly with CIGR and CSBE/SCGAB Proceedings, Québec City, QC, Canada, 13–16 June 2010; p. 1.

García-Ruiz, J.M.; López-Bermúdez, F.; Jordán, A. The effects of s\oil erosion and sediment transport on soil fertility and plant productivity. Agriculture 2017, 7, 119.

Gikas, P.; Angelakis, A.N. Water resources management in Crete and in the Aegean Islands, with emphasis on the utilization of non-conventional water sources. Desalination 2009, 248, 1049–1064.

Grieves, M.; Vickers, J. Digital Twin: Mitigating Unpredictable, Undesirable Emergent Behavior in Complex Systems. In Transdisciplinary Perspectives on Complex Systems; Springer: Cham, Switzerland, 2017; pp. 85–113.

Gupta, A.; Rayeen, F.; Mishra, R.; Tripathi, M.; Pathak, N. Nanotechnology applications in sustainable agriculture: An emerging eco-friendly approach. Pant Nano Biol. 2023, 4, 100033.

Incrocci, L.; Massa, D.; Pardossi, A. New trends in the fertigation management of irrigated vegetable crops. Horticulturae 2017, 3, 37.

Incrocci, L.; Thompson, R.B.; Fernandez-Fernandez, M.D.; De Pascale, S.; Pardossi, A.; Stanghellini, C.; Rouphael, Y.; Gallardo, M. Irrigation management of European greenhouse vegetable crops. Agric. Water Manag. 2020, 242, 106393.

Jiménez, B.; Asano, T. Water Reuse: An International Survey of Current Practice, Issues and Needs; IWA Publishing: London, UK, 2008.

Jones, H.G. Stomatal control of photosynthesis and transpiration. J. Exp. Bot. 1998, 49, 387–398.

Kaburuan, E.R.; Jayadi, R. A Design of IoT-based Monitoring System for Intelligence Indoor Micro-Climate Horticulture Farming in Indonesia. Procedia Comput. Sci. 2019, 157, 459–464.

Khan, S.; Purohit, A.; Vadsaria, N. Hydroponics: Current and future state of the art in farming. J. Plant Nutr. 2020, 44, 1515–1538.

Khormizi, H.Z.; Malamiri, H.R.G.; Ferreira, C.S.S. Estimation of Evaporation and Drought Stress of Pistachio Plant Using UAV Multispectral Images and a Surface Energy Balance Approach. Horticulturae 2024, 10, 515.

Kötke, D.; Gandrass, J.; Bento, C.P.M.; Ferreira, C.S.S.; Ferreira, A.J.D. Occurrence and environmental risk assessment of pharmaceuticals in the Mondego River (Portugal). Helyion 2024, 10, e34825.

Koukounaras, A. Advanced greenhouse horticulture: New technologies and cultivation practices. Horticulturae 2020, 7, 1.

Lamm, F.R.; Stone, K.; Dukes, M.; Howell, T.; Robbins, J.; Mecham, B. Emerging technologies for sustainable irrigation: Selected papers from the 2015 ASABE and IA irrigation symposium. Trans. ASABE 2015, 59, 155–161.

Leão, T.; Costa, B.; Bufon, V.; Aragón, F. Using time domain reflectometry to estimate water content of three soil orders under savanna in Brazil. Geoderma Reg. 2020, 21, e00280.

Lee, S.K.; Lee, J.H. Effect of hydroponic nutrient solution concentration on the growth and yield of cucumber in a plant factory system. Hortic. Environ. Biotechnol. 2015, 56, 33–39.

Leitão, I.A.; Van Schaik, L.; Iwasaki, S.; Ferreira, A.J.D.; Geissen, V. Accumulation of airborne microplastics on leaves of different tree species in the urban environment. Sci. Total Environ. 2024, 948, 174907.

Li, Y.; Liu, P.; Li, B. Water and fertilizer integration intelligent control system of tomato based on internet of things. In Proceedings of the Cloud Computing and Security: 4th International Conference, ICCCS 2018, Haikou, China, 8–10 June 2018; Revised Selected Papers, Part VI 4. Springer International Publishing: Berlin/Heidelberg, Germany, 2018; pp. 209–220.

Minhas, P.S.; Ramos, T.B.; Ben-Gal, A.; Pereira, L.S. Coping with salinity in irrigated agriculture: Crop evapotranspiration and water management issues. Agric. Water. Manag. 2020, 227, 105832.

Mir, R.; Reynolds, M.; Pinto, F.; Khan, M.; Bhat, M. High-throughput phenotyping for crop improvement in the genomics era. Plant Sci. 2019, 282, 60–72.

Murrell, K.A.; Teehan, P.D.; Dorman, F.L. Determination of contaminants of emerging concern and their transformation products in treated-wastewater irrigated soil and corn. Chemosphere 2021, 281, 130735.

O'Neill, M.P.; Dobrowolski, J.P. Water and agriculture in a changing climate. HortScience 2011, 46, 155–157.

Orgaz, F.; Fernández, M.; Bonachela, S.; Gallardo, M.; Fereres, E. Evapotranspiration of horticultural crops in an unheated plastic greenhouse. Agric. Water Manag. 2005, 72, 81–96.

Oron, G.; DeMalach, J.; Hoffman, Z.; Cibotaru, R. Subsurface microirrigation with effluent. J. Irrig. Drain. Eng. 1991, 117, 25–36.

Pardossi, A.; Incrocci, L. Traditional and new approaches to irrigation scheduling in vegetable crops. HortTechnology 2011, 21, 309–313.

Pomoni, D.I.; Koukou, M.K.; Vrachopoulos, M.G.; Vasiliadis, L. A review of hydroponics and conventional agriculture based on energy and water consumption, environmental impact, and land use. Energies 2023, 16, 1690.

Regulation (EU) 2020/741, "Regulation (EU) 2020/741 of the European Parliament and of the Council of 25 May 2020 on Minimum Requirements for Water Reuse. Off. J. Eur. Union 2019, 177, 32–55. Available online: https://eur- lex.europa.eu/legal-content/EN/TXT/PDF/?uri=CELEX:32020R0741 (accessed on 21 March 2024).

Restuccia, R. Quick Guide: Soil Moisture Sensors. 2021. Available online: https://jainsusa.com/blog/quick-guide-soil-moisture-sensors/ (accessed on 18 March 2024).

S. Mitigation of temperature, drought and viral diseases stress in vegetable crops. Int. J. Biosci. 2020, 16, 164–172.

Scharwies, J.D.; Dinneny, J.R. Water transport, perception, and response in plants. J. Plant Res. 2019, 132, 311–324.

Sevik, H.; Cetin, M. Effects of Water Stress on Seed Germination for Select Landscape Plants. Pol. J. Environ. Stud. 2015, 24, 689–693. [Google Scholar] [CrossRef] [PubMed]

Singh, D.; Biswal, A.; Samanta, D.; Singh, V.; Kadry, S.; Khan, A.; Nam, Y. Smart high-yield tomato cultivation: Precision irrigation system using the Internet of Things. Front. Plant Sci. 2023, 14, 1239594.

Strock, J.S.; Dell, C.J.; Schmidt, J.P. Drainage water management for water quality protection. J. Soil Water Conserv. 2007, 62, 144A–153A.

Wahab, A.; Muhammad, M.; Munir, A.; Abdi, G.; Zaman, W.; Ayaz, A.; Khizar, C.; Reddy, S.P.P. Role of Arbuscular Mycorrhizal Fungi in Regulating Growth, Enhancing Productivity, and Potentially Influencing Ecosystems under Abiotic and Biotic Stresses. Plants 2023, 12, 3102.

Wang, S.; Xu, J. Excessive Water and Drainage Management in Agriculture: Disaster, Facilities Operation and Pollution Control. Water 2022, 14, 2500.

Xudayev, I.; Fazliev, J.S.; Ayusupova, A. Water saving up-to-date irrigation technologies. IOP Conf. Ser. Earth Environ. Sci. 2021, 868, 12–14.

Yalin, D.; Craddock, H.A.; Assouline, S.; Mordechay, E.B.; Ben-Gal, A.; Bernstein, N.; Chaudhry, R.M.; Chefetz, B.; Fatta-Kassinos, D.; Gawlik, B.M.; *et al.* Mitigating risks and maximizing sustainability of treated wastewater reuse for irrigation. Water Res. X 2023, 21, 100203.

Yang, L.; Xia, L.; Zeng, Y.; Han, Q.; Zhang, S. Grafting enhances plants drought resistance: Current understanding, mechanisms, and future perspectives. Front. Plant Sci. 2022, 13, 1015317.

Yao, S.; Merwin, I.A.; Bird, G.W.; Abawi, G.S.; Thies, J.E. Orchard floor management practices that build soil quality and improve tree performance. HortScience 2005, 40, 2101–2106.

Zhang, F.; Zhang, Y.; Weidang, L.; Gao, Y.; Gong, Y.; Cao, J. 6G-Enabled Smart Agriculture: A Review and Prospect. Electronics 2022, 11, 2845.

Zhang, M.; Han, Y.; Li, D.; Xu, S.; Huang, Y. Smart Horticulture as an Emerging Interdisciplinary Field Combining Novel Solutions: Past Development, Current Challenges, and Future Perspectives. Hortic. Plant J. 2023.

Zinkernagel, J.; Maestre-Valero, J.F.; Seresti, S.Y.; Intrigliolo, D.S. New technologies and practical approaches to improve irrigation management of open field vegetable crops. Agric. Water Manag. 2020, 242, 106404.

Zolghadr-Asli, B.; McIntyre, N.; Djordjevic, S.; Farmani, R.; Pagliero, L. The sustainability of desalination as a remedy to the water crisis in the agriculture sector: An analysis from the climate-water-energy-food nexus perspective. Agric. Water Manag. 2023, 286, 108407.

11

Role of Sustainable Approaches in Management of Chickpea Pod Borer (*Helicoverpa armigera* Hubner) Population in Chickpea Crop

K.K. Maurya

Department of Agriculture Zoology and Entomolog, R.B.S. College, Bichpuri Campus. Agra, Uttar Pradesh

Abstract

Chickpea (Cicer arietinum L.) is the third most important pulse crop grown all over the world. Chickpea is infested by an average of about 60 insects-pests, of which gram pod borer, Helicoverpa armigera (Hubner) (Lepidoptera : Noctuidae) is known to be the key pest. Helicoverpa is responsible for causing up to 90 % damage in chickpea due to its regular occurrences. In order to manage this problem, growers are tempted to increase the amounts of pesticides, but indiscriminate or injudicious use of pesticides has resulted in residues in the food chain, pesticide resistance, and pest resurgence, in addition to causing harm to non-targeted beneficial organisms and the environment. Here, we reviewed the sustain- able approaches to reduce the incidence of pod borer and achieve sustainability in chickpea production systems through the adoption of an integrated approach involving host plant resistance, good agronomic practices, and judicious use of chemical and biological methods.

Keywords: *Helicoverpa armigera, Chickpea, IPM, Sustainable, Management, Resistant*

Introduction

Pulses are important food crops, which provide an in expensive source of protein in vegetarian diet in India. India is one of the largest producers of grain legumes and probably grows a large number of varieties of pulses than any other country. According to Indian Council of Medical Research (ICMR) optimum per capita requirement of the pulses per day to maintain normal health is 104 g (Anonymous, 2013). However, not even half of this quantity is available to the people of India.

Among the pulses, Chickpea (*Cicer arietinum* Linnaeus) is one of the prominent crops sown in rabi season. Chickpea (*Cicer arietinum* L.) is the third most important pulse crop in the worlds, with a cultivated area of 14.84 million hectares, a production of 15.08 million tons, and an average yield of 1.01 t/ha in 2020 (FAOSTAT, 2021)., which is significantly less than the estimated potential of 6 t/ha under favorable growing conditions (Thudi *et al.*, 2016). Chickpea is the third most widely grown pulse crop with global production of 15 million tons after dry beans (27 million tons) and dry pea (16 million tons) (FAO-2020). Chickpea is the main legumes crop in our diet of consumes from different parts of the world, mostly in Africa and Asian countries (Sofi *et al.*, 2020b). Chickpea is cultivated mainly in arid and semi-arid areas in more than 50 countries across the Mediterranean basin, Central Asia, East Africa, Europe, Australia, and North and South America (Dadon *et al.*, 2017). Chickpea (*Cicer arietinum* L.) is the most important legumes crop in India and is considered as "King of pulses". Chickpea is one of the most grown and consumed pulse in India as well as in worlds and it is traditionally commercialized as seeds, flour or canned food.

Chickpea (*Cicer arietinum* L.) is the premier pulse crop of India, grown all over the country mainly Madhya Pradesh, Rajasthan, Uttar Pradesh, Maharashtra, Karnataka and Haryana states under Rabi season. It is hardy, deep rooted dry land crop, able to withstand unfavorable conditions of the environment that would prove fatal for most other crops. Chickpea is a good source of essential amino acids such as tryptophan, methionine and cystiene and is the primary source of high quality protein for the largely vegetarian population of India and for those who live under the poverty line.

The constraints in increasing the production of chickpea include lack of knowledge among the farmers who cannot routinely apply fertilizers and pesticides. Chickpea yield is unstable because of its susceptibility to several biotic and a-biotic stresses including insects, diseases and drought. The accurate monitory, forecasting and density dependent factors of pest population of insect is the corner stone for economically viable, environmentally safe and socially acceptable for practices by the farmers; are essential for decision making by the plant protection specialists The emergence and occurrence of the pest over a specific space and time are dependent on climatic or geographical distribution, plant phenology and existence of natural enemies in the cropping system of the region. Hence, studies have been undertaken on population dynamics of *H. armigera* in chickpea crop so as to find out its infestation at different phenological stages for this purpose, management operation according to adoption of forecasting modules weather factors viz., temperature, relative humidity, wind velocity, evaporation rate have been timed to correlate along with population fluctuations to find out the probable a-biotic factors its multiplication. Despite of pod borer's importance, no concerted efforts have so far been made to study its economic threshold and economic injury level which are imperative and indispensable components for exploitation of effective management of the pest. The attempts have also been made to work out these parameters along with the considerations on protection techniques for avoiding adverse side ill effects of insecticides by incorporation of safer methods like microbiological agents combination with insecticides in the Central plain zone of Uttar Pradesh against, this notorious pest on chickpea

Insect pests are the main constraints which limit the production of chickpea. Various factors responsible for low production and productivity of chickpea are poor genetic base, weeds, diseases and insect pests. Major insect-pests of chickpea are cutworm (*Agrotis ipsilon* Hufnagel), gram pod borer (*Helicoverpa armigera* Hubner), gram semilooper (*Autographa nigrisigna* Walker), aphid (*Aphis craccivora* Koch) and tur pod bug (*Clavigralla gibbosa* Spinola) (Sithanantham *et al.*, 1983). Among these, *H. armigera* is most important and major cause limiting its productivity in the country.

There are two different types of chickpea grown in different part of counties i.e. Kabuli and Desi. It's different from each other based on the size, shape, and colour of seeds. Nutritionally, it contains 24 % protein, 59 % carbohydrates and 3.2 % minerals (Bakr *et al.*, 2004). Chickpea contains 18-29 % protein, 4-7% lipids and 5-60 % starch (Espinosa-Ramiraz and Serna –Saldivar, 2019., Safi *et al.*, 2020b).

Injury and Losses by *Helicoverpa armigera* on Chickpea

Chickpea attacked by more than 60 insects- pests' right from germination to maturity and also in storage (Shrivastava *et al.*, 2003). Gram pod borer (*Helicoverpa armigera* (Hubner) Hardwick (Lepidoptera: Noctuidae) is a cosmopolitan, polyphagous and notorious pest which infest more than 182 plants species. It infests the crop during early stages and continues up to flowering, podding and till the crop maturity (Bhagwat *et al.*, 2000). *Helicoverpa* is major pest in most part of the country (Yadav *et al.*, 2021). Eleven different insect-pests attacks and causes damage to chickpea crop. Among these, *Helicoverpa armigera* (Hubner), is considered to be the most devastating insect-pest (Anwar and Shafique 1993). *Helocoverpa armigera* causes damage to pods of chickpea is about 30-40 % (Luckmann and Metcaf 1975; Saleem and Younis 1982; Rahman 1990; Hashmi 1994), which may be increases to 80-90% in conducive environments (Sehgal and Ujagir 1990; Sachan and Katti 1994). The pest feeds voraciously from seedling stage to maturity and causes about 50-60 percent damage to chickpea pods (Khare *et al.*, 1977). In India *Helicoverpa armigera* casues 70-95% damage to chickpea crop from germination to maturing (Prakash *et al.*, 2007).

Pod borer, *Helicoverpa armigera* (Hubner) (Lepidoptera: Noctuidae) is the most prominent insect species that causes major economic damage to this crop. It is highly polyphagous pest attacks over 182 plants species including both widely grown and economically important crops as cotton, maize, tobacco, pigeon pea, chickpea and tomato etc. The yield loss in chickpea due to pod borer was reported as 10-60 per cent in normal weather conditions, while it was 50-100 per cent in favorable weather conditions, particularly in the states where frequent rains and cloudy weather are prevailing during the crop season. The yield loss in chickpea due to pod borer was reported as 30-80% in normal weather condition. The single caterpillar of *Helicoverpa armigera* may destroy 30-40 pods before maturity (Ram & Agrawal, 2007).

There are several reports which showed that *H. armigera* has developed resistance to all the major insecticide classes and it has become increasingly difficult to control its population in India, because of a combined effect of decreased sensitivity to acetylcholinesterase, higher levels of esterases, phosphatases and the expression of P-glycoprotein in resistance larvae.

Biology of *Helicoverpa armigera*

Helicoverpa armigera (Hubner) is the most devastating pest with high mobility and migratory potential and distributed all over the country (Forrester *et al.*, 1993; Wakil *et al.*, 2010). The spherical yellowish eggs laid by female of *Helicoverpa armigra* on tender parts, buds and flowers of the chickpea plants. The eggs oh *Helicoverpa armigera* were laid singly during night because of the moth having nocturnal behavior. The freshly laid eggs are yellowish white and glistering at first, later changed to dark brown before hatching (Ali *et al.*, 2009).

The oviposition period lasts for 5 to 24 days and female lay about 3000 eggs in night on leaves flowers and pods. The incubation period is varies from 2 to 5 days and the variation is depends on weathers parameters like temperature and humidity. The larval period depend not only on the temperature but also on the nature and the quality of the host plant. The larvae emerge after 3.37±0.09 days of egg incubation. The duration of larval development varies on different crop like 15.2 days on maize and 23.8 days on tomato. The total numbers of larval instars varies from 5 to 7. The first and second larval instars are yellowish-white to reddish-brown to blackhead capsule (Ali *et al.*, 2009). Whereas fully grown larvae are straw-yellow to green, pink or light brown to reddish- brown with lateral brown strips, and the head as well as prothoracic legs, are dark brown to black (Cunningham *et al.*, 1999; Zalucki *et al.*,1994).

The pupae are obtect type with mahogany-brown colour upto 19 mm in length, smooth and rounded on both anteriorly and posteriorly. The pupation takes place in soil, clods, or fallen materials on the soil in chickpea crop. The pre-pupal period lasts for 1 to 4 days. The larvae spin a loose web of silk before pupation. In non- diapausing pupae, the pupal period ranges from about 6 days at 35 °C to over 30 days at 15 °C (Ali *et al.*, 2009). The diapausing period for pupae can lasts several months. Pale coloured adults are produced from pupae exposed to temperature exceeding 30 °C. The female moth lives longer than males. In laboratory, longevity varies from 1 to 23 days for male and 5 to 28 days for females (Pearson 1958). The pod borer moth is large and brown with a V-shaped speck and dull black border on hind wing. Life cycle of *Helicoverpa* armigera take about 30-34 days with an average temperature o 28 °C form egg to adult (Zalucki *et al.*, 1986)

The pod borer exhibits facultative diapauses, which allow it to survive adverse weather conditions in both winter and summer seasons. The winter diapauses are induced by the exposure of larvae to short photoperiods and low temperatures. In china and India, pod borer populations are composed of tropical, sub-tropical and temperate ecotype. In subtropical Australia, the pod borer undergoes diapauses during the winter, when the temperatures are low. High temperature can also induce diapauses. It enter a true summer diapauses when the larvae are exposed to very high temperatures(43°C for 8 hours daily), although the proportion of females enterning diapauses is nearly half compared with that of males. At these temperatures non-diapausing males are sterile.

Host Plant and Biology of *Helicoverpa armigera*

The insect *Helicoverpa armigera* Hubner is a wide spread species. The *Helicoverpa armigera* caterpillars feed on 300 plants species, including agricultural commercially important crops, and causes more than $ 2 billion of losses to agriculture in the world (Anonymous, 2013; Ahmad *et al.*, 2016). Among biotic factors chickpea is infested by nearly 60 insects species in which *Agrotis ipsilon* (Ratt.), *Helicoverpa armigera* (Hub.), *Autagrapha nigrisigma* (Walk.) and *Aphis craccivora* (Koch.) are the pest of major importance

(Acharjee & Sharma, 2013). Among these the major damage caused by gram pod borer which is polyphagous in nature. *Helicoverpa armigera* is one of the serious pest of chickpea, which feed more than 150 crops through the world (Vinutha *et al.*, 2013).

Helicoverpa armigera is a polyphagous and voracious feeder of diverse plant species. Over 200 plant species including economically important crops such as; Cotton, Sorghum, pigeonpea , chickpea, maize, soybean, tomato, pepper, bean, peas, sunflowers, niger seeds and many other horticultural crops were preferred by H.armigera (Fite *et al.*, 2018, Tabkew, 2004; Waktole,1996).

The intensity of damage depends on the types of host plants on which they feed. In chickpea, adult females laid the eggs on leaves, flowers and young pods. After hatching of eggs the neonate larvae feed on the tenders and succulents parts of the host plants. The first, second and third instar larvae initially feed on the foliage (young leaves) of chickpea and a few other legumes but mostly on the flower buds of cotton, pigeon pea etc. Larvae shift from foliage feeders to developing seeds and fruits as larval instar development progresses (Reed and Pawar 1982). The insects larvae destroys various plant parts like pods, buds, flowers and fruits of its host plants preferring the harvestable part of economically important crops throughout the world (Sarwar, 2013).

The pod borer attacks crops from seedling to maturity, damaging all parts of the plants (leaves, flowers and pods). Initially, the larvae feed on the leaves and tender twigs of the chickpea plant, and later, when the chickpea pods are formed, the larvae bores and feed inside.

The young pod borer larvae (first, second and third instar) occasionally enter the pod and feed upon the developing chickpea grains, but more often they feed from outside the pod with only the anterior part of its body in the pod (Saxena 1978; Singh & Singh 1987). Patel and his coworkers 2010 reported that pod borer caused maximum entry holes at the basal region of chickpea pods irrespective of the genotype

Management strategies for *Helicoverpa armigera*

Management of several of the important insect-pests of crops is becoming increasingly difficult owing to a number of factors by which pest have been able to offset the effect of control interventions. Several of the present day pest management problems may be ascribed to the excessive and indiscriminate use of chemical pesticides during the last about three decades. Routine application of pesticides for controlling insect-pests, without taking into consideration whether economic damage to the crop is being inflicted by the pest or not, has resulted into several problems like pest resistance, pest resurgence, pest replacement, killing natural enemies of pests, presence of toxic residue in food, and soil and environmental pollution, etc., apart from increasing in cost of plant protection. Entomologists have been trying to find solution to these problems through integrated pest management of which monitoring of insect activity through regular surveys is an important component. In facts the knowledge of insect activity is indispensable since pest management system cannot operate without accurate estimates of pests and natural enemies' population or without reliable assessment of plant damage and its effect on yield.

1. Monitoring of *Helicoverpa* Through Pheromone Traps

Regularly monitoring the key pest is a vital component of any Integrated Pest Management (IPM) program that helps to take control measures. An effective control strategy always depends on accurate monitoring of damaging stages of the in- sect. Monitoring or recording is also necessary to understand the major factors influencing pest population to forecast its incidence. Pheromone traps can be incorporated to develop predictive models designed to provide information on probable oviposition patterns, and population abundance of pod borer moth catch is positively correlated with the larval count (Prabhakar *et al.*, 1998). Pheromone traps are described as a good tool to monitor lepidopterous pests (Malik & Ali, 2002). Monitoring of *H. armigera* by using pheromone traps or light traps was successfully implemented and incorporated as basic tools used to monitor, forecast, prediction, and control deci- sions based upon the populations of the moth catch (Fite *et al.*, 2020a, 2020b; Yenumula & Prabhakar, 2012).

2. Agronomical Manipulations

Sowing time

Sowing chickpea at the optimum time one of the most important factors affecting crop yield. Weather parameters like temperature, humidity and sunshine played an important role in regulating pod borer populations. It was reported that the pod borer larval population positively correlated with temperature, whereas relative humidity and rainfall inhibit the larval population (Kumar and Bisht 2013; Shinde *et al.*, 2013). The crop sown later suffered most from the pod borer infestation, as compared with that which was sown earlier. Early sowing of chickpea resulted in low pod borer larval population and pod damage percentage in India (Garg 1990; Choudhary *et al.*, 2015; Parmar *et al.*, 2015). The yield of chickpea grains was also decreased when sowing was delayed, It indicating that the sowing time is direct correlation with pest incidence (Borah 1998; Singh *et al.*, 2002). Grow early maturing crops to avoid dominant *Helicoverpa* armigera populations late in the season (Fakrudin *et al.*, 2003).

Plant Density

Higher plant density and planting geometry likely favors pest growth by creating a micro-climate conductive to their dark-loving *Helicoverpa* larvae. Plant density also affects the extent of damage caused by pod borer. In general, higher plant density favors increased pod damaged (Reed *et al.*, 1987; Naresh *et al.*, 1987; Quadeer and Singh 1989; Begum *et al.*, 1992). However, higher plant density may not necessarily result in yield loss due to compensation in total pod number per unit urea (Sithanantham and Reed 1979; Pimbert 1990). Population dynamics of *Helicoverpa* armigera on chickpea revealed that higher population of larvae and pupae and higher density crop harbored more population than low plant density crops (Sithanantham *et al.*, 1981; Kant *et al.*, 2007). Increasing in seed rate results in closure planting geometry which support higher *Helicoverpa* larval populations.

Increasing the seed rate from 75-100 kg per h greatly increased the percentage of pod damage. Further increasing the seed rate from 100-125 kg per h did not show a significant increase in pod damage. But increase pod damaged in higher seed rate did not affect chickpea grain yield due to the higher plant population per unit area, which compensated for the increased pod damage (Anil *et al.*, 2011,).

Nutrient Management

Fertilizers are primary applied to produce high yield of a crop, but their use may have indirect effect on pest attack. This effect may be positive or negative (Coaker 1987). The indiscriminate use of NPK makes a plant vigorous, bushy, succulents, rendering them more susceptible to pod borer. The bushy nature of plants provide better shelter to dark loving pod borer, causing increased pod damage (Hussain *et al.*, 2009). Low to moderate doses of NPK fertilizers led chickpea to have lower pod borer infestation, but higher doses of NPK led to higher pod borer infestation (Hussain *et al.*, 2009). Applying organic manures like farm yard manures, neem cake and vermicompost resulted in the lowest pest population compared to the application of inorganic fertilizers (Rao 2003; Muddukumar 2007; Singh and Singh 2007).

Cropping System/Intercropping/Mixed Cropping

Cropping systems of chickpea with certain crops has been shown to reduce damage from pod borer. This may be a result of the companion crop harboring higher no of natural enemies or non-preference for egg laying by pod borer in a field containing the intercrop. Intercropping generally delayed the appearance of major chickpea pests and reduced their incidence, particularly the linseed intercrop (Borah *et al.*, 2010). Tripathi and coworkers (2008) found that the minimum larval population and highest chickpea grain yield were found in chickpea + mustard, followed by chicpea + barley and chickpea + wheat. Similar results have also been supported by Reena *et al.*, (2009) and Hossain (2003). Coriander and other nectar- rich plants encourage parasitoids activity. Chickpea intercrop with coriander harbored the minimum *Helicoverpa* armigera population and reduced the incidence of pod borer (Singh and Pandey 2014; Chandra *et al.*, 2014). The importance of intercropping of chickpea with sunflower in minimize the incidence of *Helicoverpa* was also supported by Patter *et al.*, 2012

Trap Crop

The basic strategy in the cultivation of trap crops which is preferred host of pest for feeding, shelter and breeding over a small area to attract and kill the pest. But the area of trap crop should not be more than one tenth of the main crop area. Okra is preferred host plant of *Helicoverpa,* cotton jassid, spotted boll worm. The infested okra fruits that have a large population of *Helicoverpa* should be removed and destroyed periodically (Thimmaiah & Raju 1991). Marigold and Nicotiana rustica grown as trap crops are also preferred hosts of *Helicoverpa.*

Bird Perches

Several species of insectivorous birds have been found to feed on crop insect-pests, including pod borer (Chakravarthy 1988), which have been known to reduce the larval population to the extent of 84 percent in Punjab, India. Among the predatory birds, black drongo, house sparrow, blue jays, cattle egret, rosy pastor, and mynah are common predators on a large number of *H. armigera* and lepidopteran insets of chickpea, pigeonpea, and groundnut crops (Gokhle and Ametha 1991).

3. Host Plant Resistance

Host plant resistance present an ideal means combating notorious insect pests including *H. armigera* (Gola *et al.*, 2018; Xiaoyi *et al.*, 2015). Farmers believe that use of local landrace varieties; mainly desi chickpea varieties are more moderately resistance to *Helicoverpa armigera* than the kabuli types (Fite *et al.*, 2019). Genotype resistance to *Helicoverpa armigera* accumulated more oxalic acid on the leaves and growth inhibition on *H. armigera* larvae when include in a semi-artificial diet (Mitsuru *et al.*, 1995).

4. Biological Control

It is an important tool in integrated pest management programs. In this type of control the populations H.armigera manage by using natural enemies and microbes. It is a sustainable approach to manage insect pest population without any harmful impact on natural phenomenon. Biological control by using natural enemies and microbes are important components of IPM programs against plant pests (Barratt *et al.*, 2018; Van Lenteren *et al.*, 2018) including H.armigera (Seid & Tebkew, 2002; Crowder *et al.*, 2010; Fite *et al.*, 2020a, 2020b).

5. Microbial Pesticides (Bacteria, Fungi and Virus Based Insecticides)

Microbial Some of the microbes effective for the control of *H. armigera* larvae have included bacteria; Baci lus thuringinensis (Bt.) (Das *et al.*, 2016), viruses; nuclear poly-hedrosis virus (NPV) (Gomez Valderrama *et al.*, 2018) and entomopathogenic fungi (Fite *et al.*, 2019). Baci lus thuringiensis is available as a selective spray that only kills moth larvae and the pathogen is extensively studied on various caterpillars including *H. armigera* (Das *et al.*, 2016; Silva *et al.*, 2018, Fite *et al.*, 2019). Baci lus thuringiensis Berliner variety kurstaki were effective at the rate of 1.50 kg/ha was successfully controlled the first and second instar larvae of *H. armigera* up to an average of 99.58 and 90.42 %, respectively (Alemayehu *et al.*, 1993). *Helicoverpa* armigera larvae were reportd to be extremely prone to Bt. delta toxin (Li and Bouwer, 2012). Entomopathogenic fungi mainly trains of *Beauveria bassiana* and *M. anisoplae* have been reported to be effective against *H. armigera* (Jarrahi and Safavi, 2016; Klavandi *et al.*, 2018).

6. Plant and Animal Based Extracts

Plant materials are known by farmers to be environmentally benign, safer, and more cost-effective compared to synthetic pesticides (Kamanula *et al.* 2011), as well as difficult to adulterate when produced or harvested by farmers themselves. The most well-known and commonly used plant extract is azadirachtin, isolated from the seed, wood, bark, leaves, and fruits of the neem tree (*Azadirachta indica*). Azadirachtin has both antifeedent and growth-retarding properties and can lead to death at any stage in the life cycle, probably by interfering with the neuroendocrine control of metamorphosis in insects (Roy and Dureja 1998). Neem and garlic extract have larvi- cidal, toxic, repellent, ovicidal, antifeedent and antioviposition effects on insect-pests (Zhu *et al.*, 2001). Applying Neem Seed Kernel Extract (NSKE 5%) treatment reduced the pod borer population in chickpea (Gupta 2007; Pachundkar *et al.* 2013; Hussain *et al.*, 2016). Leaf, bark, and seed extract from *Annona squamosa* have pesticidal and insect antifeedent properties (Bisen and Bansal 2014). Mishra *et al.*, (2013) reported a significant decrease in the percentage of pod damage after spraying vermin-wash with neem oil and custard apple leaf extract.

Insecticide Resistance and Integrated Management for *H. armigera*

The indiscriminate use of synthetic insecticides has led to several adverse effects including resistance developments in *H. argimera*. Hence, due to its economic importance, insecticides are frequently used for the controlling of *H. argimera* which has been developed resistance to several conventional insecticides from the organo- phosphate, pyrethroid, and carbamate groups. Moreover, these insecticides are harmful to beneficial insects, thus it has become important to use insecticides that are ecologically safe for natural enemies. In modern days, several new insecticide groups having new chemistries with safer molecules viz., Diamide, Oxadiazine, Phenyl Pyrazol, Pyrazol, Pyrrole, Ketoenol, Furanicotinyl, Tetronic Acid Derivatives, Nicotinoids, Sulphite ester acaricides, Carbazate acaricides and Insect Growth Regulators have been discovered and commercialized for uses in crop protection. Most of these groups of pesticides play an important role in managing many pests with good bio-efficacy & low mammalian toxicity, which make them attractive replacement for synthetic organic pesticides or conventional insecticides. These novel groups of pesticides are likely to play an important role in Integrated Pest Management programme in future. The ability of these new groups of insecticide to be effective at low doses, high level of selectivity, greater specificity to target pests along with low safe to environment, helpful in resurgence and resistant management and replaced many conventional compounds. Some chemical group of insecticides such as spinosad (chemi- cal group: Spinosyn) and indoxacarb is safe to the natural enemies (Nasreen *et al.*, 2003; Williams *et al.*, 2003). Spinosad was reported to be very effective against the larvae of *H. argimera* both on contact and by ingestion (Carneiro *et al.*, 2014; Hamed & Khan, 2003). Recently, the application of Indoxacarb (Avaunt 150 SC) at 0.3 Lha^{-1} or Spinosad (Tracer 480 SC) at 0.15 Lha^{-1} three times with a week interval were reported to be effective in reducing the percentage of pod damage, mean lar- vae per plant and as a result increased grain yield ha^{-1} under field condition and can be advised for the management of *H. armigera* (Mihretie *et al.*, 2020). However, still investigations for very soft and effective chemical group of insecticides are remained as research outlook. There- fore, these insecticides can be used in the Integrated Pest Management (IPM) program for the control of *H. armigera*. Nowadays, IPM is being used to find ecologically sound, economically viable, and environmentally safe ways of pest management and received better attention. Single pest management could not provide sustainable and effective control; therefore all the available options have to be incorporated together whenever they are compatible. For instance, by integrating neem, *Helicoverpa* Nuclear Polyhedrosis Virus (HaNPV) and endosulfan resulted in reducing damaged pods, achieved the highest grain yield in chickpea (Visalakshmi *et al.*, 2005). In another study conducted by Kumari *et al.* (2015) a combination of pheromone trap, Bt and HaNPV were significantly reduced damaged pods. Further, research should focus on IPM of all major pests emphasizing eco- nomically important crops.

Conclusion

Production and productivity of chickpea in India were constrained mainly by *H. armigera.* The economic losses inflicted by a single insect pest couldn't be achieved through a single pest management method. Therefore, a comprehensive systems approach is needed to help the small scale farmers whose livelihoods have been either directly or indirectly dependent upon agriculture. Various control measures such as; cultural, monitoring, botanical pesticides, the use of natural enemies, host plant resistance, synthetic chemicals, and IPM strategies have been suggested under this review. IPM encompasses all pest management techniques, which should consider farmers' economic gain/ profit, environment, sustainability, effectiveness, holistic and systematic approach.

References

Ahmad, R. and Chandel, S. F. (2004). Farmers field evaluation of IPM module against *Helicoverpa armigera* infesting chickpea. Archives of Phytopathology and Plant Protection, 37(2): 133-137.

Ahmad, S., Ansari, M.S., Siddiqui, M.H. and Hussain, M. (2016). Effect of intercrops on management of *Helicoverpa armigera* in chick pea agro-ecosystem. Ann. Pl. Protec. Sci. 24(2): 208-212

Alemayehu, R., Tadesse, G. M., & Mengistu, H. (1993). Laboratory and field evaluations of *Bacilus thuringiensis* Berliner var. kurstaki on African boll- worm, *Helicoverpa* armigera (Hübner). In Proceeding of the first annual conference of crop protection society of Ethiopia, Addis Ababa, Ethiopia (pp. 28).

Ali, A., Choudhury, R. A., Ahmad, Z., Rahman, F., Farmanur, R. K., Syed, K. A., & Ahmad, S. K. (2009). Some biological characteristics of *Helicoverpa armigera* on chickpea. Tunisian Journal Plant Protection, 4(1), 99–106.

Anil kumar R, Nandan B, Sharma JP, Kumar J (2011). Effect of phosphorus and seed rate on growth and productivity of bold seeded Kabuli chickpea in subtropical Kandi areas of Jammu and Kashmir. PlantArchives 10(1):125–129.

Anonymous (2013). Annual report 2012–13. Indian Institute of Pulses Research (IIPR), Kanpur.

Anonymous (2013). Directorate of Economics and Statistics, Department of Agriculture and Cooperation. Govt. of India. pp: 85-87.

Anonymous (2019). Project Co-ordinators Report 2018-19. All Indian Co-ordinated Research project on chickpea, ICAR-Indian Institute of Pulses Research. pp.1- 42.

Anwar M, Shafique M (1993). Integrated control of gram pod borer, *Helicoverpa armigera* (Hubner) in Sindh. In: Proceedings of Pakistan Congress of Zoology, pp 215–222.

Avilla, C., Vargas-Osuna, E., González-Cabrera, J., Ferré, J., & González- Zamora, J. E. (2005). Toxicity of several delta-endotoxins of *Bacilus thuringiensis* against *Helicoverpa armigera* (Lepidoptera: Noctuidae) from Spain. Journal of Invertebrate Pathology, 90, 51–54.

Bakr MA, Afzal MA, Hamid A, Haque MM, Aktar MS (2004). Blackgram in Bangladesh. Lentil, Blackgram and Mungbean Development Pilot Project, Pulses Research Centre, BARI, Gazipur.

Bajya, D. R., Ranjith, M., & Raza, S. K. (2015). Evaluation of Beauveria bassiana against chickpea pod borer, *Helicoverpa armigera* and its safety to natural enemies. Indian Journal of Agricultural Science, 85(3), 378–381.

Bar-El Dadon, S.; Abbo, S.; Reifen, R. (2017). Leveraging traditional crops for better nutrition and health—The case of chickpea. Trends Food Sci. Technol; 64, 39–47.

Barratt, B. I., Moran, V. C., Bigler, F., & van Lenteren, J. C. (2018). The status of biological control and recommendations for improving uptake for the future. Bio Control, 63, 155–167.

Begg, G. S., Cook, S. M., Dye, R., Ferrante, M., Franck, P., Lavigne, C., & Lövei, G. L. (2017). A functional overview of conservation biological control. Crop Protection, 97, 145–158.

Begum N, Husain M, Chowdhury SI (1992). Effect of sowing date and planting density on pod borer incidence and grain yield of chickpea in Bangladesh. Int. Chickpea News let 27:19–21.

Bhagwat, V.R. and Sharma, S.B. Evaluation of chickpea genotypes for resistance to legume pod borer (*Helicoverpa armigera*) and root- knot nematode (*Meloidogyne javanica*). Indian Journal of Plant Protection, 28 (1):69-73. (2000)

Bhatt NJ, Patel RK (2001). Screening of chickpea cultivars for their resistance to gram pod borer *Helicoverpa armigera* (Hb). IndianJEnt. 63:277–280.

Bheemaraya, V. Rachappa, Raju Teggelli, Suhas Yelshetty and Y.S. Amaresh (2019). Biological activity of pod borer, *Helicoverpa* armigera (Hubner) influenced by chickpea genotypes, LegumeResearch, 42 (3) : 421-425.

B.L. Jat, K.K. Dahiya, S.S. Yadav, S. Mandhania (2021). Morpho Physico-Chemical Components of Resistance to Pod Borer, *Helicoverpa armigera* (Hübner) in Pigeonpea *Cajanus cajan* (L.) Millspaugh] Legume Research- An International Journal, 44(8): 967-976.

Borah BK, Debnath MC, Sharma KK, Das B (2010). Effect of inter-cropping on incidence of gram pod borer, *Helicoverpa armigera* Hubner in chickpea. Insect Environ 15(4):8–9.

Borah RK (1998). Influence of sowing dates on the infestation of *Helicoverpa armigera* and grain yield of chickpea (*Cicerarietinum*) in the hill zone of Assam. IndianJEnt 60(3):416–417.

Chakravarthy AK (1988). Bird predators of pod borers of field bean (Lablab niger). Trop. PestMgmt 34:395–398. doi:10.1080/ 09670878809371285.

Chandra S (1987). Production technology from all India pulse improvement project in Nepal. Coordination of grain legume research in Asia. Summary proceedings of the Review and Planning Meeting for Asian Regional Research on Grain Legumes (groundnut, chick-pea and pigeon pea), 16–18 December 1985 ICRISAT, Patancheru, India, pp 45–49.

Chandra BKN, Rai PS (1974). Two new ornamental host plants of *Heliothis armigera* Hubner in India. Indian J Horti 31(2):198

Chaturvedi SK, Gurha SN, Sewak S, Ahmed R, Dikshit HK, Bahduoria P (1998). Possible combined resistance against Fusarium wilt and pod borer in chickpea (*Cicerarietinum* L.) Indian J Pulses Res 11:117– 119.

Choudhary OM, Anwala R, Sharma MM (2015). Studies on vari*et al* screening and date of sowing of chickpea [*Cicerarietinum* (L.)] against *Helicoverpa armigera* (Hub.) JEco-friendlyAgric 10(1): 58–61.

Coaker TH (1987). Cultural methods. In: Burn AJ, Coaker TH, Jepson PC (eds) The crop in integrated pest management. Academic Press, London, pp 69–88.

Crowder, D. W., Northfield, T. D., Strand, M. R., & Snyder, W. E. (2010). Organic agriculture promotes even natural enemies and natural pest control. Nature, 466, 109–112.

Cunningham, J. P., Zalucki, M. P., & West, S. A. (1999). Learning in *Helicoverpa armigera* (Lepidoptera: Noctuidae): A new look at the behaviour and control of a polyphagous pest. Bulletin Entomological Research, 89, 201–207.

Damte, T., Ali, K., Tesfaye, A., & Chichaybelu, M. (2018). Progresses in insect pest management research in highland food legumes of Ethiopia. Ethiopian Journal of Crop Science, 6(3), 343–368.

Damte, T., & Mitiku, G. (2020). Pattern of egg distribution by Adzuki bean bee- tle, Ca losobruchus chinensis (L.) (Coleoptera: Chysomelidae) in stored chickpea under natural infestation. Journal of Stored Products Research, 88, 101683.

Damte, T., & Mitiku, G. (2021). Evaluation of traditional method of stored product protection: Effect of mixing tef (Eragrostis tef) grains with stored chickpea on occurrence of Adzuki bean beetle (*Callosobruchus chinensis*) and its natural enemies. International Journal of Pest Management.

Danso, J. K., Osekre, E. A., Opit, G. P., Manu, N., Armstrong, P., Arthur, F. H., Camp- bell, J. F., Mbata, G., & McNeill, S. G. (2018). Post-harvest insect infestation and mycotoxin levels in maize markets in the Middle Belt of Ghana. Journal of Stored Products Research, 77, 9–15.

Das, A., Datta, S., Sujayanand, G. K., Kumar, M., Singh, A. K., Shukla, A. A., Ansari, J., Kumar, M., Faruqui, L., Thakur, Sh., Kumar, P. A., & Singh, N. P. (2016). Expression of chimeric Bt gene, Cry1Aabc in transgenic pigeon (cv. Asha) confers resistance to *Helicoverpa armigera* Hubner. Plant Cell, Tis- sue and Organ Culture, 127(3), 705–715.

Deshmukh, S. G., Sureja, B. V., Jethva, D. M., & Chatar, V. P. (2010). Field efficacy of different insecticides against *Helicoverpa armigera* (Hubner) infest- ing chickpea. LegumeResearch, 33(4), 269–273.

Dhruv H Antala, DR Patel and LL Makvana (2022). Evaluation of insecticides against *Helicoverpa armigera* Hubner in chickpea, Journal of Entomology and Zoology Studies; 10(5): 115-117.

D. Sagar, Suresh M. Nebapure, Subhash Chander (2022). Assessment of *Helicoverpa* armigera (Hubner) Adult Population in Chickpea using Weather Based Indices. Legume Research-An International Journal, Volume 45 Issue 2:260-264

Espinosa-Ramiraz J. and Serna –Saldivar, S.O (2019). Wet milled chickpea co- product as a alternative to obtain protein isolates ,LWT, 115, 10

FAOSTAT Rome (2021). https://www.fao.org/faostat/es/#data/QCL

FAOSTAT (FAO) (2019). http://www.fao.org/faostat/en/#data/QC

FAO (2020). FAOSTAT food and agriculture data

Fikre, A., Desmae, H., & Ahmed, S. (2020). Tapping the economic potential of chickpea in Sub-Saharan Africa. Agronomy, 10, 1707.

Fite, T., Damte, T., Tefera, T., & Negeri, M. (2020a). Evaluation of commercial trap types and lures on the population dynamics of *Helicoverpa armigera* (Hubner) (Lepidoptera: Noctuidae) and its effects on non-targets insects. Cogent Food and Agriculture, 6, 1–1771116. parasitoid species of the cotton bollworm, *Helicoverpa armigera* (Hubner) (Lepidoptera: Noctuidae) in chickpea growing districts of Ethiopia. Biocontrol Science and Technology, 31(5), 541–545.

Fite, T., Tefera, T., Negeri, M., & Damte, T. (2020b). Effect of Azadirachta indica and Mi letia ferruginea extracts against *Helicoverpa armigera* (Hubner) (Lepidoptera: Noctuidae) infestation management in chickpea. Cogent Food & Agriculture. https://doi.org/10.1080/23311932.2020.1712145

Fite, T., Tefera, T., Negeri, M., Damte, T., & Sori, W. (2018). Management of *Helicoverpa* armigera (Lepidoptera: Noctuidae) by nutritional indices and botanical extracts of *Milettia ferruginea* and *Azadirachta indica*. Advances in Entomology, 6, 235–255.

Fite, T., Tefera, T., Negeri, M., Damte, T., & Sori, W. (2019). Evaluation of *Beauveria bassiana*, Metarhizium anisopliae, and Baci lus thuringiensis for the man- agement of *Helicoverpa armigera* (Hubner) (Lepidoptera: Noctuidae) under laboratory and field conditions. Biocontrol Science and Technology, 30(3), 278–295.

Forrester, N. W., Cahill, M., Bird, L. J., & Layland, J. K. (1993). Management of pyre- throid and endosulfan resistance in *Helicoverpa armigera* (Lepidoptera: Noctuidae) in Australia. Bu letin of Entomological Research, 1, 1–132.

Garg DK (1990). Effect of sowing date on the incidence of *Helicoverpa* armigera (Hub) and yield of chickpea in Kumaon Hills, India. Int Chickpea Newslet 23:14–15.

G.K. Taggar, Ravinder Singh, H.S. Randhawa, H.K. Cheema (2021) .Novel Insecticides for the Management of Pod Borer Complex in Pigeonpea Crop. Legume Research-An International Journal, Volume 44 Issue 10: 1179-1185.

Gokhale VG, Ameta OP (1991). Predatory behavious of house sparrow, Passer domesticus L. in the population regulation of Heliothis sp. infesting chickpea, *Cicerarietinum*. Indian J Ent 53:631–634.

Gómez Valderrama, J. A., Barrera, G., López-Ferber, M., Belaich, M., Ghiringhelli,P. D., & Villamizar, L. (2018). Potential of betabaculo viruses to control the tomato leafminer, Tutaabsoluta (Meyrick). Journal of Applied Entomology, 142, 67–77.

Gowda CLL, Lateef SS, Smithson JB, Reed W (1983). Breeding for resistance to *Heliothis armigera* in chickpea. In: Proceedings of the National Seminar on Breeding Crop Plants for resistance to pests and diseases, 25–27 May 1983, School of Genetics, Tamil Nadu Agricultural University Coimbatore, Tamil Nadu, India.

Gupta MP (2007) Management of gram pod borer, *Helicoverpa armigera* (Hubner) in chickpea with biorationals. Nat Prod Radiance 16(5): 391–397.

Hashmi AA (1994) Insects pest management, taxonomy and identification. Pakistan Agricultural Research Council (PARC), Islamabad, p 1034.

Hossain, A. M., Azizul, H. M., & Prodhana, M. Z. (2008). Incidence and damage severity of *Helicoverpa armigera* (Hübner) in chickpea (*Cicer Arietinum* L.). BangladeshJ.Sci.Ind. Res, 44(2), 221–224.

Hussain, D., Muhammad, S., Ghulam, G., & Muneer, A. (2015). Resistance in a field population of *Helicoverpa armigera* (Hubner) (Lepidoptera: Noctui- dae). Journal of Entomological Science, 50(2), 119–128.

Hossain MA, Prodhan MZH, Haque MA (2009) Response of NPK fertilizer on incidence of pod borer, *Helicoverpa armigera* (Hubner) and grain yield of chickpea. Bangladesh J Sci Ind Res 44(1):117– 124. doi:10.3329/bjsir.v44i1.2720.

Hossain MA, Haque MA, Ahmad M, Prodhan MZH (2010) Development of an integrated management approach for pod borer, *Helicoverpa armigera* (Hubner) on chickpea. Bangladesh J. Agric. Res. 35(2):201–206. doi:10.3329/bjar.v35i2.5882.

Shah MJ, Hussain K, Ahmad B, Khan AQ (2016) Integrated management of *Helicoverpa armigera* on different genotypes of Kabuli chickpea in Punjab, Pakistan. Int. J. Biosci. 9(2):110– 119. doi:10.12692/ijb/9.2.110-119.

Jarrahi, A., & Safavi, S. A. (2016). Fitness costs to *Helicoverpa armigera* after exposure to sub-lethal concentrations of *Metarhizium anisopliae* sensu lato: Study on F1 generation. Journal Invertebrate Pathology, 138, 50–56.

Kalvnadi, E., Mirmoayedi, A., Alizadeh, M., & Pourian, H. (2018). Sub-lethal concentrations of the entomopathogenic fungus, Beauveria bassiana increase fitness costs of *Helicoverpa armigera* (Lepidoptera: Noctuidae) offspring. Journal of Invertebrate Pathology, 158, 32–42.

Kamanula J, Sileshi GW, Belmain SR, Sola P, Mvumi BM, Nyirenda GKC, Nyirenda SPN, Stevenson PC (2011). Farmers' insect-pest management practices and pesticidal plant use for protection of stored maize and beans in Southern Africa. Int. J. Pest. Mgmt. 57(1): 41–49. doi:10.1080/09670874.2010.522264.

Kant K, Kanaujia KR, Kanaujia S (2007). Role of plant density and abiotic factors on population dynamics of *Helicoverpa armigera* (Hübner) in chickpea. Annals of Plant Protection Sciences.

Keshav Mehra, Veer Singh (2022). Impact of Intercrop Combinations in the Management of Gram Pod Borer, *Helicoverpa armigera* (Hubner) in Chickpea. Legume Research-AnInternationalJournal, Volume 45 Issue 7: 926-929.

Khare BP, Ujagir R. Protection of pulse crops from insect pest ravages. Indian Farming Digest. 1977;10(2):31-35.

Kumari, K., Anil, K., Saha, T., Goswami, T. N., & Singh, S. N. (2015). Bio-intensive management of *Helicoverpa armigera* (Hübner) in chickpea. Journal of Eco-Friendly Agriculture, 10(1), 50–52.

Li, H., & Bouwer, G. (2012). Toxicity of Baci lus thuringiensis Cry proteins to *Helicoverpa armigera* (Lepidoptera: Noctuidae) in South Africa. Journal of Invertebrate Pathology, 109, 110–116. New York, pp 3–35.

M.A. Waseem, Sobita Simon and Sasya Thakur (2017). Effect of Intercropping on the Incidence of Gram Pod Borer, *Helicoverpa armigera* in Chickpea. Int. J. Curr. Microbiol.App.Sci ; 6(10): 2619-2624.

Malik, M. F., & Ali, L. (2002). Monitoring and control of codling moth, Cydia pomone la (Lepidoptera: Tortricidae) by pheromone traps in Quetta, Pakistan. Asian Journal of Plant Science, 1, 201–202.

Mihretie, A., Yimer, D., Wudu, E., & Kassaw, A. (2020). Efficacy of insecticides against African bollworm (*Helicoverpa armigera* Hubner) on chickpea (*Cicer arietinum*) in the lowlands of Wollo, Norteastern Ethiopia. Cogent Food & Agriculture, 6(1), 1833818.

Mishra K, Singh K, Tripathi CPM (2013). Management of pod borer (*Helicoverpa armigera*) infestation and productivity enhancement of gram crop (*Cicer arietinum*) through vermin-wash with bio-pesticides. World J Agric Sci 9(5):401–408. doi:10.5829/idosi. wjas. 2013.9.5.1749.

Mitsuru, Y., Susan, E., Cowgil, L., & John, A. W. (1995). Mechanism of resistance to *Helicoverpa armigera* (Lepidoptera: Noctuidae) in chickpea: Role of Oxalic acid in leaf exudate as an antibiotic factor. Journal of Economic Entomology, 88(6), 1783–1786.

Muddukumar (2007). Organic nutrient management and plant protection in green chickpea production system. M.Sc. (Agri.) Thesis, University of Agricultural Sciences, Dharwad, India.

Mumtaz Hussain1, Khawaja Shafique Ahmad, Muhammad Majeed, Ansar Mehmood, Abdul Hamid, Malik Muhammad Yousaf, Muhammad Shafiq Chaudhry, Muhammad Jahangir Shah, Khadim Hussain, Bashir Ahmad, Abdul Qadir Khan. Integrated management of *Helicoverpa armigera* on different genotypes of Kabuli chickpea in Punjab, Pakistan. Int.J.Biosci. 2016. Vol. 9, No. 2, p. 110-119.

Nasreen, A., Mustafa, G., & Ashfaq, M. (2003). Selectivity of some insecticides to Pakistan Journal of Biological Sciences, 6(6), 536–538.

Niedermayer, S., Pollmann, M., & Steidle, J. L. (2016). Lariophagus distinguendus (Hymenoptera: Pteromalidae) (Förster)-past, present, and future: The history of a biological control method using L.distinguendus against different storage pests. Insects, 7, 39.

Pachundkar N, Kamble P, Patil P, Gagare P (2013). Management of gram pod borer *H. armigera* in chickpea with neem seed kernel extract as a natural pest management practice in Bhojdari.

Pandey R, Ujagir R (2008). Effect of intercropping on *Helicoverpa armigera* (Hub.) infesting chickpea. Ann Pl Protec Sci 16(2):320– 324.

Parmar SK, Thakur AS, Marabi RS (2015). Effect of sowing dates and weather parameters on the incidence of *Helicoverpa armigera* (Hubner) in chickpea. The Bioscan 10(1):93–96.

Patel IS, Patel PS, Patel JK, Acharya S (2010). Pod boring pattern of gram pod borer, *Helicoverpa armigera* Hub. in chickpea varieties. Insect Environ 16(1):41–42.

Pattar PS, Mansur CP, Alagundagi SC, Karbantanal SS (2012). Effect of intercropping systems on gram pod borer *Helicoverpa armigera* Hubner and its natural enemies in chickpea. Indian J. Ent. 74(2): 136–141.

Pawar VM, Aleemuddin M, Bhole BB (1987). Bio-efficacy of HNPV in comparison with Endosulfan against pod borer on chickpea. Int Chickpea Newslet 16:4–6.

Pearson EO (1958). Insect pests of cotton in tropical Africa.

Pimbert MP (1990). Some future research directions for Integrated Pest Management in chickpea: a viewpoint. Chickpea in the nineties. In : Proceeding of Second International Workshop on Chickpea Improvement, 4–8 December 1989, ICRISAT, Patancheru, India, pp 151–163.

Prabhakar M, Singh Y, Newpane SP (1998). Predicting *Helicoverpa armigera* of chickpea by using sex pheromone traps. Indian J Ent 60(4):339–334.

Prakash, M. R., Ram, U., & Tariq, A. (2007). Evaluation of chickpea (*Cicer arietinum* L.) germplasm for the resistance to gram pod borer, *Helicoverpa armigera*

Hubner (Lepidoptera: Noctuidae). Journal of Entomology Research, 31, 215–218.

Prasad D, Chand P (1989). Effect of intercropping on the incidence of *Helicoverpa armigera* (Hub.) and grain yields of chickpea. J. Res. Birsa Agric. Univ. 1(1):15–18.

Prasad D, Kumar B (2002). Impact of intercropping and Endosulfan on the incidence of gram pod borer infesting chickpea. Indian J. Ent. 64: 405–410.

Rahman MM (1990). Infestation and yield loss in chickpea due to pod borer in Bangladesh. Bangladesh J. Agric. Res. 15(2):16–23.Rahman MM, Mannan MA, Islam MA (1982) Pest survey of major summer and winter pulses in Bangladesh. In the Proceedings of the National Workshop on Pulses. Bangladesh Agricultural Research Institute, Joydebpur, Dhaka, pp 265–273.

Rao KR (2003). Influence of host plant nutrition on the incidence of *Spodoptera litura* and *H. armigera* on groundnut. IndiaJEnt 65(3): 386–392.

Reed W (1983). Estimation of crop losses due to insect-pests in pulses. IndianJEnt 2:263–267.

Reed W, Pawar CS (1982). Heliothis: a global problem. In: Reed W, Kumble V (eds) Proceedings of International Workshop on *Heliothis* Management, 15–20 November 1981. ICRISAT, Patancheru, pp 9–14.

Reed W, Cardona C, Sithanantham S, Lateef SS (1987). Chickpea insect- pests and their control. In: Saxena M, Singh KB (eds) The chickpea. CAB International, Wallingford, pp 283–318.

Reena, Singh SK, Sinha BK, Jamwal BS (2009). Management of gram pod borer, *Helicoverpa armigera* (Hubner) by intercropping and monitoring through pheromone traps in chickpea. Karnataka J Agric Sci 22(3-special issue):524–526.

Roy NK, Dureja P (1998). New eco-friendly pesticides for integrated pest management. Pesticides World, New Delhi, pp 16–22.

Sachan JN, Katti G (1994) Integrated pest management. In: Proceedings of International Symposium on Pulses Research, 2–6 April 1984, IARI, New Delhi, India, pp 23–30.

Safi., S.A., Singh J., Muzaffar, K. Majid, D. & Dar, B.N. (2020b). Physiochemical characteristics of protein isolates from native and germinated chickpea cultivars & their noodle quality. International Journals of Gastronomy and Food Science, 222, 100258

Saleem M, Younis A (1982). Host plants and nature and extent of damage of *Helicoverpa.*

Saxena HP (1978). Pests of grain legumes and their control in India. In: Singh SR, van Emden HF, Taylor TA (eds) Pest of grain legumes: ecology and control. AcademicPress,London, pp 15–23.

Sehgal VK, Ujagir R (1990). Effect of synthetic pyrethroids, neem extracts and other insecticides for the control of pod damage by *Helicoverpa armigera* (Hübner) on chickpea and pod damage yield relationship at Pantnagar in Northern India. Crop Prot 9:29–32. doi: 10.1016/0261-2194(90)90042-6.

Seid, A., & Tebkew D. (2002). African bollworm: A key pest in pulses production in Ethiopia: Status and Need. Proceedings of the National Workshop held at the Plant Protection Research Center Ambo, Ethiopia.

Sekar PR, Vankataaiah M, Rao NV, Rao VR, Rao VSP (1996). Monitoring of insect resistance in *Helicoverpa* arimigera (Hub.) from area receiving heavy insecticide application in Andhra Pradesh. J.Ent.Res. 20(2):93–102.

Sergey Semerenk, and Nadezhda Bushneva (2020). The efficiency of control of the number of *Helicoverpa armigera* Hbn. in sunflower sowings. E3S Web of Conferences 222, 0; DAIC2020 203.

Silva, I. F., Baldin, E. L., Specht, A., Sosagómez, D. R., Roque-Specht, V. R., Morando, R., & Paula-Moraes, S. V. (2018). Biotic potential and life table of *Helicoverpa armigera* (Hübner) (Lepidoptera: Noctuidae) from three Brazilian regions. Neotropical Entomology, 47(3), 344–351.

Shafiq Ur Rehman1, Muhammad Waqar Hassan1, Moazzam Jamil1 and Muhammad Akram (2017). Screening of some chickpea (*Cicer arietinum* L.) varieties against *Helicoverpa armigera* under semi arid climatic conditions. Legume Research, 40 (2): 379-383.

Singh, B. D. (2002). Plant breeding: Principles and methods. Kalyani Publishers.

Shinde YA, Patel BR, Mulekar VG (2013). Seasonal incidence of gram caterpillar, *Helicoverpa armigera* (Hub.) in chickpea. Current Biotica 7:79–82.

Singh K, Pandey J (2014). Evaluation of inter-cropping for management of pod borer, *Helicoverpa armigera* Hub. in chickpea. Bioinfolet 11(2c):688–691.

Singh RN, Singh J (1987). Pattern of boring of chickpea pods by *Heliothis armigera* Hubn. Indian J.Agric.Sci. 57(5):376–377.

Sithanantham S, Reed W (1979). Plant density and pest damage in chick- pea. Int Chickpea Newslet 1:9–10.

Sithanantham S, Tahhan O, Hariri G, Reed W (1981). The impact of winter sown chickpeas on insect-pest and their management. In: Saxena MC, Singh KB (eds) Proceedings of Workshop on Ascochyta Blight and Winter Sowing of Chickpeas. ICARDA, Aleppo, pp 179–187.

Srivastava SK (2003). Relative preference of different chickpea genotypes by *Helicoverpa armigera* (Hubner) and estimation of yield losses under late sown condition. Indian Journal of Pulses Research. 2003;16(2):144-146.

Su Htet San, D. Sagar, Vinay Kumari Kalia, Veda Krishnan (2022). Effect of Different Chickpea Genotypes and Its Biochemical Constituents on Biological Attributes of *Helicoverpa armigera* (Hubner). Legume Research-An International Journal, Volume 45 Issue 4: 514-520.

Tebkew, D., & Mohamed, D. (2006). Chickpea, lentil and grass pea insect pest research in Ethiopia: A review. P. 260–273. In Ali, K., Keneni, G., Ahmed, S., Malhotra, R., Beniwal, S., Makkouk, K., & Halila, M. H. (eds.) Food and forage legumes of Ethiopia: Progress and prospects. Proceedings of a Work- shop on Food and Forage Legumes (pp. 22–26) Sept 2003, Addis Ababa, Ethiopia. ICARDA, Aleppo, Syria.

Tebkew, D., & Ojiewo, C. O. (2017). Incidence and within field dispersion pattern of pod borer, *Helicoverpa armigera* (Lepidoptera: Noctuidae) in chickpea in Ethiopia. Archives of Phytopathology and Plant Protection, 50, 17–18.

Thimmaiah, G and Raju G.T.T. 1991. Possibilties of using okra (Bhindi) as a trap crop in cotton boll worm management. Curr.Ent.Res. USA 20 (7) 140-142.

Thudi, M.; Chitikineni, A.; Liu, X.; He, W.; Roorkiwal, M.; Yang, W.; Jian, J.; Doddamani, D.; Gaur, P.; Rathore, A.; *et al.* Recent breeding programs enhanced genetic diversity in both desi and kabuli varieties of chickpea (*Cicer arietinum* L.). Sci. Rep. 2016, 6, 38636.

Tripathi SR, Sharma SK (1984) Extent of damage and incidence of *Heliothis armigera* (Hubner) (Noctuidae: Lepidoptera) on different varieties of gram in Terai belt of eastern Uttar Pradesh, India. Ann Ent 2:31–35.

Tripathi A, Sharma RC, Dwivedi PK, Pandey M (2008). Effect of intercropping and insecticide on pod borer incidence in chickpea. J Food Legumes 21(3):187–188.

Ujagir R, Khere BP (1987). Preliminary screening of chickpea genotypes for susceptibility to *Heliothis armigera* (Hub.) at Pantnagar, India. Int Chickpea Newslet 17:14.

Vikrant, Dharm Raj Singh, Sanjeev Kumar, Kaushal Kishor, Ram Kewal (2020). Bio- efficacy of Insecticides against *Helicoverpa armigera* in Chickpea. Legume Research-An International Journal, Volume 43 Issue 2: 276-282.

Van Lenteren, J. C., Bolckmans, K., Kohl, J., Ravensberg, W. J., & Urbaneja, A. (2018). Biological control using invertebrates and microorganisms: plenty of new opportunities. BioControl, 63, 39–59.

Visalakshmi, V., Ranga, G. V., & Arjuna, R. P. (2005). Integrated pest management strategy against *Helicoverpa armigera* in chickpea. Indian Journal of Plant Protection, 33(1), 17–22.

Wakil, W., Ghazanfar, M. U., Kwon, Y. J., Qayyum, M. A., & Nasir, F. (2010). Distribu- tion of *Helicoverpa armigera* Hübner (Lepidoptera: Noctuidae) in tomato fields and its relationship to weather factors. Entomological Research, 40(6), 290–297.

Waktole, S. (1996). Managements of cotton insect pests in Ethiopia: A review. In Proceedings of integrating biological control and host plant resistance. Addis Ababa,Ethiopia CTA/ IAR/IIBC (pp. 224–252).

Williams, T., Valle, J., & Vinuela, E. (2003). Is the naturally derived insecticide Spinosad compatible with insect natural enemies? Biocontrol Science and Technology, 13(5), 459–475.

Xiaoyi, W., Haifeng, C., Aihua, S., Rong, Z., Zhihui, S., Xiaojuan, Z., *et al.* (2015). Laboratory testing and molecular analysis of the resistance of wild and cultivated soybeans to cotton bollworm, *Helicoverpa armigera* (Hübner). The Crop Journal, 3, 19–28.

Yadav, A., Keval, R. and Yadav, A. (2021). Evaluation of insecticides and biopesticides against *Helicoverpa armigera* in different modules of short duration pigeonpea. Legume Research. DOI: 10.18805/LR-4313.

Zahid MA, Islam MM, Reza MH, Prodhan MHZ, Rumana Begum M (2008). Determination of economic injury levels of *Helicoverpa armigera* (Hubner) in chick pea. Bangladesh J Agric Res 33(4):555–563. doi:10.3329/bjar. v33i4.2288.

Zalucki MP, Daglish G, Firemponeng S, Twine PH (1986). The biology and ecology of *Heliothis armigera* (Hübner) and *H. punctigera* (Wallengren) (Lepidoptera: Noctuidae) in Australia: what do we know? Aust J Zool, 42:329–346.

Zalucki, M. P., Murray, D. A., Gregg, P. C., Fitt, G. P., Twine, P. H., & Jones, C. (1994). Ecology of *Helicoverpa armigera* (Hübner) and *Helicoverpa punctigera* (Wallengren) in the inland of Australia: Larval sampling and host plant relationships during winter and spring. Australian Journal of Zoology, 42, 329–346.

Index

A
Abiotic Stress: 76, 87
Agrochemicals: 2, 150, 155, 156, 157, 158
Agronomy: 49, 86, 87, 90, 111, 133, 182
Amino Acids: 22, 26, 27, 108, 170
Antagonist: 22, 144
Antibacterial: 17, 20, 23, 106, 107, 110, 112, 161
Antifungal: 14, 29, 104, 157
Antimicrobial: 9, 17, 20, 22, 35, 37, 105, 107, 110, 111, 150, 151, 157, 159
Arthropods: 45
Azadirachtin: 6, 10, 11, 22, 178

B
Bacillus Thuringiensis: 4
Bacillus: 4, 14, 15, 140, 149, 151
Bacterial Blight: 77, 86, 106, 160
Bactericide: 162
Beauveria: 4, 178, 181, 183, 184
Biocontrol Agents: 24, 141, 145, 150, 151, 153, 158, 159
Biocontrol: 24, 141, 144, 145, 149, 150, 151, 153, 158, 159, 182, 183, 187
Biofertilizer: 148
Biological Control: 13, 129, 132, 133, 142, 149, 150, 178, 181, 185, 187
Biomass: 30, 31, 43, 96
Biopesticide: 4
Biopolymer: 163
Biosensor: 34, 37, 109
Botanical: 1, 2, 5, 7, 10, 34, 180, 183
Botany: 10

C
Callose: 14, 152
Carbon Sequestration: 138, 141
Caterpillar: 4, 44, 59, 61, 172, 186
Chemical Control: 56, 73, 132
Chitosan: 15, 20, 22, 105, 106, 110, 111, 112, 152, 157, 160, 163
Chlorophyll: 46
Chlorosis: 22, 114
Climatic Factors: 146
Contact Insecticide: 8
Crop Damage: 18, 51, 96, 131, 133
Crop Disease: 91, 95
Crop Loss: 101, 123, 134, 147
Crop Protection: 2, 3, 21, 32, 33, 36, 38, 55, 87, 145, 148, 179, 180, 181
Crop Rotation: 29, 30, 37, 103, 140, 143, 148
Cuticle: 13

D
Defoliation: 114
Degradation: 5, 21, 99, 104, 106, 108, 138, 144
Disease Cycle: 150, 153
Disease Forecasting: 150, 153
Disease Management: 1, 17, 18, 53, 95, 101, 103, 104, 105, 107, 109, 111, 113, 114, 117, 118, 119, 135, 145, 146, 148, 150, 153, 155, 156, 161
Disease Resistance: 13, 14, 24, 25, 27, 28, 29, 31, 34, 35, 36, 39, 73, 84, 87, 88, 90, 143, 157
DSRNA: 21, 84, 164

E
Ecosystem: 16, 43, 59, 109, 138, 139, 143, 180
Elevated Co2: 13
Entomology: 10, 89, 133, 134, 135, 182, 183, 184, 185, 186
Entomopathogenic: 14, 142, 144, 178, 184
Enzyme: 6, 26, 35, 108, 109, 112, 138
Epidemic: 70, 123, 150
Epidemiology: 153
Essential Nutrients: 29, 143
Essential Oils: 8, 9, 10, 11, 160
Ethylene: 14, 163
Evaporation: 5, 141, 166, 171

F
Farming Systems: 143
Fertilizer: 34, 36, 37, 49, 106, 110, 166, 183
Field Scouting: 52
Field Trials: 106
Food Chain: 18, 109, 169
Formulation: 6, 7, 14, 15, 18, 22, 34, 105, 106, 107, 159
Fungal Pathogen: 104, 123
Fungicide: 33, 144, 149, 152, 162
Germination: 29, 33, 109, 114, 167, 171, 172
Green Pesticides: 1, 2, 3, 4, 5, 7, 8, 9, 10, 11
Herbicide: 18, 32, 33, 34, 35, 38, 105, 112
Host-Pathogen: 29, 109
Humidity: 46, 51, 127, 132, 155, 171, 172, 175
Hydrolysis: 107

I
Immune Response: 14, 19
Insect Resistance: 2, 28, 102, 186
Insect Vector: 129
Insecticide: 4, 7, 8, 106, 107, 172, 179, 186, 187
Integrated Pest Management: 33, 54, 56, 89, 97, 114, 131, 133, 145, 148, 160, 175, 178, 179, 182, 185, 186, 187
Intercropping: 29, 30, 176, 177, 184, 185, 186, 187
IPM: 33, 54, 55, 89, 97, 114, 130, 131, 132, 133, 145, 148, 149, 150, 153, 160, 162, 169, 175, 178, 179, 180
ISR: 13, 14, 163

L
Leaf Blight: 20, 25, 26, 28, 31, 35, 38, 63, 64, 66, 72, 86, 87, 122, 123
Leaf Spot: 26, 63, 72, 89, 160
Lignification: 19, 31
Livelihood: 2, 121

M
Macronutrients: 17, 161
Manure: 29, 35, 139, 140, 148, 162
Microbial: 1, 3, 10, 13, 14, 15, 16, 29, 30, 31, 36, 112, 137, 138, 139, 140, 141, 143, 144, 160, 162, 178
Microbiome: 16, 139
Microflora: 16, 148
Mode of Action: 8, 108
Molecular Biology: 54
Molecular Marker: 76
Mosaic Virus: 21, 22, 77, 88, 104, 105, 135
Mycelium: 112
Mycorrhiza: 139, 148

N
Nano: 17, 18, 19, 20, 21, 22, 34, 35, 37, 38, 101, 102, 103, 104, 105, 106, 107, 108, 109, 110, 111, 112, 165
Nanoparticles: 17, 18, 19, 20, 21, 34, 35, 36, 37, 38, 39, 101, 102, 103, 104, 110, 111, 156, 157
Nanotechnology: 19, 20, 35, 37, 101, 102, 103, 106, 109, 110, 111, 112, 145, 156, 165
Natural Enemies: 85, 126, 129, 133, 170, 175, 176, 178, 179, 180, 181, 182, 185, 187
Necrosis: 22, 114
Neem: 6, 7, 176, 178, 179, 180, 185, 186
Nematicide: 22, 49
Nitrate: 25, 165
Nitrogen Fixation: 139
Nutrient Deficiency: 24

O
Organic Farming: 7, 145, 148, 149, 150
Organic Manure: 139
Oxidative Stress: 20, 22

P
Pathogen: 13, 14, 15, 17, 19, 22, 23, 24, 25, 26, 28, 29, 30, 31, 32, 33, 36, 43, 51, 53, 73, 90, 101, 102, 103, 104, 108, 109, 123, 143, 146, 147, 150, 153, 154, 155, 156, 157, 160, 161, 162, 163, 178
Pathogenicity: 163, 164
Pathology: 1, 31, 34, 35, 37, 41, 55, 88, 101, 111, 135, 181, 184
Pest Resurgence: 169, 174
Pest: 1, 2, 3, 4, 5, 6, 7, 8, 9, 10, 32, 33, 34, 35, 43, 44, 45, 48, 50, 51, 52, 53, 54, 55, 56, 57, 59, 60, 67, 68, 71, 73, 84, 85, 86, 89, 91, 93, 95, 96, 97, 98, 99, 101, 102, 103, 104, 107, 110, 113, 114, 115, 117, 118, 119, 121, 122, 125, 126, 127, 128, 129, 130, 131, 132, 133, 134, 135, 140, 141, 142, 144, 145, 147, 148, 149, 153, 160, 162, 169, 170, 171, 172, 173, 174, 175, 176, 177, 178, 179, 180, 182, 183, 184, 185, 186, 187
Pesticide Resistance: 152, 169
Pesticide: 2, 10, 13, 15, 18, 100, 106, 110,

111, 114, 122, 150, 152, 155, 156, 158, 161, 169
Pheromone Trap: 180
Pheromone: 52, 127, 142, 175, 180, 184, 185, 186
Phloem: 31
Phosphate: 20, 25, 31, 37, 108, 162, 179
Phosphorus: 17, 24, 25, 30, 139, 162, 180
Photosynthesis: 27, 166
Phytoalexins: 29
Phytohormone: 38
Phytopathology: 35, 36, 38, 90, 119, 180, 187
Phytoplasma: 37
Phytosanitary: 147
Phytotoxicity: 7, 104
Plant Defense: 19, 20, 21, 28, 36, 152, 158, 160
Plant Extract: 8, 178
Plant Health: 17, 36, 41, 42, 43, 44, 48, 54, 96, 98, 99, 141, 142, 161
Plant Immunity: 34, 137
Plant Metabolism: 8, 27, 33
Plant Morphology: 19
Plant Nutrition: 19, 29, 50, 97, 145, 185
Plant Physiology: 26, 127
Pollution: 102, 112, 122, 152, 167, 175
Population Dynamics: 2, 33, 126, 171, 176, 182
Precipitation: 128, 130, 132
Protein Synthesis: 21, 108
Pseudomonas: 14, 15, 25, 38, 110, 150, 151, 160, 161, 163
Pyrethrum: 7

Q
Quarantine: 114, 145, 147
Quorum Sensing: 14

R
Residue: 9, 109, 144, 151, 175
Rhizobacteria: 148, 163
Rhizosphere: 21, 24, 28, 162
RNAi: 18, 21, 22, 37, 38, 84, 88, 104, 163, 164
Root Exudates: 143

S
Salicylic Acid: 13, 19, 27, 152, 162
Sar: 13, 20, 27, 29, 31, 162
Secondary Metabolites: 3, 22
Silicon: 25, 28, 34
Soil Fertility: 38, 51, 102, 137, 138, 139, 145, 146, 148, 165
Soil Health: 29, 122, 125, 137, 138, 139, 140, 141, 142, 143, 144
Soil Ph: 25, 51
Soil Salinity: 138
Soil Structure: 140, 141, 143
Soil Texture: 138, 148
Solanaceae: 18
Spore: 15, 29, 154
Spray: 7, 13, 31, 35, 37, 90, 164, 178
Stomata: 13, 109
Stress Response: 85
Sustainable: 1, 2, 3, 10, 13, 16, 21, 24, 29, 33, 34, 37, 41, 47, 54, 55, 57, 87, 91, 92, 99, 100, 101, 102, 109, 110, 111, 114, 131, 133, 137, 138, 139, 140, 143, 144, 145, 146, 147, 148, 149, 151, 153, 155, 156, 157, 158, 159, 160, 161, 163, 164, 165, 166, 167, 169, 171, 173, 175, 177, 178, 179, 180, 181, 183, 185, 187
Systemic Resistance: 13, 15, 29, 36, 37, 145, 148, 151, 162, 163
Systemic: 13, 15, 25, 27, 29, 31, 33, 36, 37, 145, 148, 151, 162, 163

T
Terpenes: 8, 152, 160
Toxicity: 2, 7, 8, 18, 28, 87, 101, 105, 158, 179, 181, 184
Transpiration: 166
Trichoderma: 14, 140, 142, 149, 150, 151

U
Ultraviolet: 4
Uptake: 8, 13, 17, 19, 24, 30, 44, 156, 162, 163, 181
Virulence: 33, 85, 88, 153, 162, 163
Volatile Oils: 8

W
Waterlogging: 142, 164
Weed Management: 36, 97
Wilt: 17, 23, 36, 38, 51, 53, 72, 123, 148, 181

X
Xylem: 26, 157
Y
Yield Loss: 68, 122, 172, 176, 185
Z
Zinc: 17, 21, 22, 27, 30, 31, 34, 38, 157, 158, 162